T0219556

Cambridge Elements ≡

Elements in Flexible and Large-Area Electronics
edited by
Ravinder Dahiya
University of Glasgow
Luigi Occhipinti
University of Cambridge

1D SEMICONDUCTING NANOSTRUCTURES FOR FLEXIBLE AND LARGE-AREA ELECTRONICS

Growth Mechanisms and Suitability

Dhayalan Shakthivel
University of Glasgow

Muhammad Ahmad
University of Surrey

Mohammad R. Alenezi
The Public Authority for Applied Education and Training, Kuwait, and University of Surrey

Ravinder Dahiya
University of Glasgow

S. Ravi P. Silva
University of Surrey

CAMBRIDGE
UNIVERSITY PRESS

CAMBRIDGE
UNIVERSITY PRESS

University Printing House, Cambridge CB2 8BS, United Kingdom

One Liberty Plaza, 20th Floor, New York, NY 10006, USA

477 Williamstown Road, Port Melbourne, VIC 3207, Australia

314–321, 3rd Floor, Plot 3, Splendor Forum, Jasola District Centre,
New Delhi – 110025, India

79 Anson Road, #06–04/06, Singapore 079906

Cambridge University Press is part of the University of Cambridge.

It furthers the University's mission by disseminating knowledge in the pursuit of
education, learning, and research at the highest international levels of excellence.

www.cambridge.org
Information on this title: www.cambridge.org/9781108724654
DOI: 10.1017/9781108642002

© Dhayalan Shakthivel, Muhammad Ahmad, Mohammad R. Alenezi, Ravinder Dahiya
and S. Ravi P. Silva 2019

First published 2019

A catalogue record for this publication is available from the British Library.

ISBN 978-1-108-72465-4 Paperback
ISSN 2398-4015 (online)
ISSN 2514-3840 (print)

1D Semiconducting Nanostructures for Flexible and Large-Area Electronics

Growth Mechanisms and Suitability

Elements in Flexible and Large-Area Electronics

DOI: 10.1017/9781108642002
First published online: October 2019

Dhayalan Shakthivel
University of Glasgow

Muhammad Ahmad
University of Surrey

Mohammad R. Alenezi
*The Public Authority for Applied Education and Training, Kuwait,
and University of Surrey*

Ravinder Dahiya
University of Glasgow

S. Ravi P. Silva
University of Surrey

Author for correspondence: Ravinder Dahiya, Ravinder.Dahiya@glasgow.ac.uk

Abstract: Semiconducting nanostructures such as nanowires (NWs) have attracted significant attention in recent years for flexible electronics applications as they offer attractive physical, chemical and optical properties. They have been used as building blocks for various types of sensors, energy storage and generation devices, electronic devices and for new manufacturing methods involving printed NWs. The response of these sensing/energy/electronic components and the new fabrication methods depends very much on the quality of NWs, and for this reason it is important to understand the growth mechanism of 1D semiconducting nanostructures. It is also important to understand the compatibility of NW growth steps and tools used in the process with the unconventional substrates, such as plastics, that are used in flexible and large-area electronics. Therefore, this Element presents at length discussion about the growth mechanisms, growth conditions and the tools used for the synthesis of NWs. Although NWs from Si, ZnO and carbon nanotubes (CNTs) are included, the discussion is generic and relevant to several other types of NWs as well as heterostructures.

Keywords: nanowires, VLS mechanism, vapour–solid mechanism, nanowire growth kinetics, silicon, carbon, ZnO, CNTs

ISBNs: 9781108724654 (PB), 9781108642002 (OC)
ISSNs: 2398-4015 (online), 2514-3840 (print)

Contents

1 Introduction

1.1 Semiconducting Nanowires (NWs)

One-dimensional (1D) or zero-dimensional (0D) nanostructures with diameters <100 nm have attracted significant interest in the past few years, as they exhibit interesting physical and chemical properties that often vary significantly from those of their bulk counterparts. For instance, the highly inert bulk gold turns out to be a reactive material in the form of 0D nanoparticles with size range of 3–50 nm [1, 2]. The enhanced physical and chemical properties of nanostructures are attributed to their high surface-to-volume ratio, change in the band gap, density of states and quantum confinement, etc. [3]. In the case of 1D nanostructures, i.e., NWs (e.g., Si NWs, ZnO NWs and carbon nanotubes (CNTs)), the two dimensions are <100 nm, which means that electrons are confined to two dimensions, leading to interesting electronic, optical, thermal and magnetic properties [4]. The prominent size-dependent effects include quantum confinement, ballistic charge transport characteristics, optical absorption and emission, metal insulator transition, etc. [5, 6]. The interesting features as a result of these size-dependent effects have been exploited to develop nanoscale FETs [7, 8], NW-based optoelectronic devices (photodiodes, LEDs, lasers and waveguides) [8], biosensors [9], gas detectors [10], energy generators (solar cells, thermoelectric and nano) [11–13] and energy storage devices [14]. The 1D morphology is most suitable for this wide range of applications, as it offers advantages in terms of single crystallinity of material, controlled chemical composition, diameter and possibility to engineer as 1D heterostructure. The latter has many variants (Figure 1.1) with huge benefits for the above-mentioned applications. Among 1D nanostructures the CNTs, having single-atomic-layer wall thickness, are the most widely explored [15]. The use of CNT in various applications depends on the conductivity (e.g., metallic or semiconductor) and the structure (e.g., single-wall or multiwall). The application potential of CNTs has also been proved with the demonstration of low-power, high-density devices for computer processors [16]. In fact, the utility of nanotubes from several materials has been demonstrated for electronics, sensors and energy devices [17, 18].

Another group of 1D nanostructures that has attracted significant interest is the heterostructure NWs (Figure 1.1). Through synthesis and subsequent assembly, they have the potential to go far beyond the limits and functionality of single-material-based 1D nanostructures. For instance, owing to the formation of an intrinsic 2D electron gas, the Ge and Si NWs with a core-shell arrangement result in transistors with improved charge carrier mobility and hence high performance [19]. NW heterostructures have also been used to

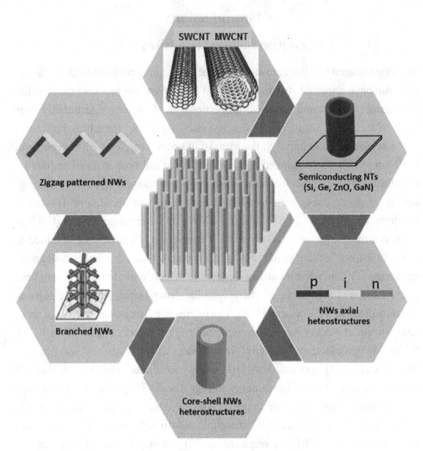

Figure 1.1 Schematic illustration of different variants of semiconducting 1D nanostructures

demonstrate large-area active matrix electronic skin [20]. The axial NW hetero-structures (Figure 1.1) are another possibility where periodic repetition of different materials with the same diameter (e.g., GaAs/GaP/GaAs …) is done along a direction [21]. These 3D hierarchical NWs have manifold increments in surface–volume ratio and have strong implications for several applications such as those discussed above. Complex NW structures such as branched NWs and zigzag NWs could be obtained by tailoring the growth conditions. Furthermore, it is possible to have heterostructures with chemically distinct building blocks to obtain unique functions for sensitive and high-performance electronic circuits over large and flexible areas.

The attractive properties of the semiconducting NWs are well suited for several applications in flexible and large-area electronics [22–30]. NWs offer several advantages over organic and inorganic thin-film materials for realising electronic devices and circuits. The intrinsic flexibility and low cost of fabrication are often stated as the merits of the organic semiconductor-based approaches for large-area electronics. The superior electronic properties of inorganic NWs (mobility~1 cm^2/V·s as compared to 1000 cm^2/V·s for single-crystal Si) give them an edge, however, when it comes to high-performance flexible electronics. Recently, by printing of single-crystal Si NWs on flexible substrates, the possibility of obtaining high-performance electronics with low-cost fabrication has also been demonstrated [20, 31–33]. By creating a suitable composite with NWs it is possible to maintain their inherent mechanical flexibility and stability while exploiting their enhanced properties, particularly when compared with inorganic thin films [34–37], organic semiconductors and 2D nanostructures (graphene, MoS$_2$, WS$_2$, etc.) [38–41]. As a result, a range of 1D nanomaterials (e.g., carbon nanotubes (CNTs), NWs of ZnO, Si, Ge, GaAs, etc.) has been explored to develop devices (e.g., Field Effect Transistors (FETs), sensors, energy harvesters and display devices, etc.) over flexible substrates [30, 33, 42–49] for many interesting solutions in wearable systems, robotic e-skin, healthcare, etc. Most of the NWs are conveniently synthesised at high temperatures. The typical temperature range (300–1000°C) to obtain high-quality NWs is not suitable for their direct growth on flexible substrates (e.g., polyimide and plastics), which usually have the low glass transition temperature (T_g) (–125°C to 350°C [50]) and thus would melt at the high temperatures needed for the growth of NWs. Therefore, there is a thermal budget issue, which calls for assessing the thermal properties of substrates commonly used in flexible electronics. This challenging issue has led researchers to explore alternative routes such as transfer and contact printing of NWs (discussed in Section 5) [30, 51–55], which decouples the high-temperature processing steps (e.g., growth of NWs or high-quality dielectric deposition) from the low-temperature processing steps (e.g., metallisation for source, drain or gate terminals of transistors) needed to obtain electronic devices after NWs [30, 48, 56, 57]. A detailed description of contact and transfer printing methods is given in [30], which is another Element published as part of this same Cambridge Elements Series.

The performance of NW-based devices, the yield of transfer process and eventually the effective utilization of 1D nanostructures are influenced by physical and chemical parameters such as diameter, crystal structure and chemical composition of NWs and the way they are synthesised [4, 58–60]. This Element focusses mainly on the growth mechanisms of 1D nanostructures – particularly for the Si NWs, ZnO NWs and CNTs, which are amongst the most researched 1D nanostructures for various applications.

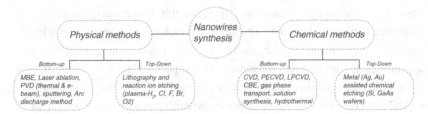

Figure 1.2 The classification of NW growth by physical and chemical methods

1.2 Nanowire Growth Methods

Wide varieties of physical and chemical methods have been developed in the past six decades for the synthesis of semiconducting NWs. As shown in Figure 1.2, these methods can be further categorised as bottom-up and top-down methods. The physical methods largely use bulk single crystalline materials (e.g., Si, Ge and GaAs wafer) to produce NWs by thinning the bulk wafer using high-energy plasma species [61, 62]. Si NWs synthesised by the e-beam lithography technique combined with dry etching of Si wafers is a representative example of the top-down or subtractive physical method. Numerous alternative chemical methods developed for the synthesis of semiconducting NWs can be classified among the bottom-up methods. Most of the chemical methods use gaseous and liquid precursors to build NWs from atoms and molecules. Among various bottom-up NW growth methods, the catalyst particle–assisted vapour–liquid–solid (VLS) growth mechanism, first reported in 1964 by R. S. Wagner for Au-catalysed Si microwhiskers [63–65], has been extensively explored for the synthesis of NWs from different materials by utilising various physical (PVD, sputtering, e-beam, MBE, etc.) and chemical vapour deposition (MOCVD and ALD) tools. A vast majority of the NW-based device prototypes for flexible electronics are based on this method [25, 49, 66, 67].

1.2.1 Catalyst Particle–Assisted VLS Growth Method

The VLS technique is a standard method used for the growth of NWs at predefined locations. The major advantages of VLS growth mechanism are (1) synthesis of single crystalline NWs suitable for a wide range of applications such as electronic devices, sensors, energy generation and storage, etc.; (2) good dimensional control allowing NWs with diameter from a few nanometres to several microns; (3) potential to tune the composition from elemental, binary to ternary compound NWs suitable for the growth of a wide variety of material systems such as Si, Ge, ZnO, CNTs, GaAs, InGaAs, etc.; (4) possibility of in situ doping during synthesis; (5) possibility of growing NWs at specified locations by suitably placing the catalyst and enabling control over wire placement, NW interspacing and length, etc.; and (6) engineering of NWs to develop novel heterostructures such as NWs with different materials in the core-shell, axial and branched heterostructured morphologies.

The VLS method uses nanosized catalyst particles for NW growth, as they help to concentrate the vapour flux at predefined locations. The size of the catalyst particles determines the diameter of the resulting NWs. Identifying a suitable catalyst particle is one of the key requirements of this process, as experimental conditions are determined by the choice of catalyst. The catalyst metal must form a liquid solution with the NW material, which could be identified using binary phase diagrams. The binary phase diagrams readily give out two critical parameters: (1) catalyst-NW material composition and (2) growth temperature where the catalyst liquid alloy and solid NW coexist. The nanoliquid surface is an ideal rough surface for the gaseous molecules' adsorption followed by feeding of atoms for NW growth. The saturated liquid containing catalyst particles leads to the crystallization of NWs of the same diameter as that of the catalyst particles.

1.2.2 Catalyst-Free Vapour–Solid Growth Method

Growth of NWs can also be achieved in the absence of a metal catalyst. Such methods involve thermally evaporating a precursor material near its melting point under a controlled environment, which leads to condensation in a self-structuring manner and hence to the growth of NWs [68–71]. Because of the lack of a liquid phase, such growth processes are called vapour–solid mechanism (VS). The arc-discharge and laser ablation methods are examples of the VS growth mechanism used for CNTs. In these methods, a solid carbon source is evaporated at high temperatures and the condensation results in the growth of MWCNTs. Similarly, Zn powder is thermally evaporated in an oxygen environment or a precursor gas is decomposed at high temperatures to yield the ZnO NWs [68]. A good control over various growth parameters using the VS methods is relatively difficult as compared with the VLS methods. The control over diameter of the NWs in VS methods is generally achieved by tuning evaporation or condensation temperatures and the vapour pressure.

This Element focusses mainly on the VLS method for 1D semiconducting nanostructures from Si, ZnO and CNTs, which are commonly used in flexible and large-area electronics. It may be noted that the mechanism does not always align with the requirements of flexible and large-area electronics, and extra steps are needed during or after the NW synthesis. For example, a range of synthesis conditions (e.g., temperature, pressure, precursor composition) needs to be controlled to obtain electronic-grade semiconducting NWs. Among these, the high-temperature (>600°C) requirement of the vapour-phase technique, needed to decompose the gaseous and liquid chemical precursors for the growth of ZnO and Si NWs, is not compatible for flexible substrates such as PET, PEN or PI, as they would melt if direct synthesis is attempted. Likewise, for CNT growth, the metal catalyst film is typically heated in the temperature range of 600–1100°C to form nanoparticles, and the hydrocarbon vapour is introduced in the reaction chamber. Growth temperatures are considerably reduced with other variants of the CVD process. For example, the photothermal CVD (PTCVD) process allows keeping substrates at lower temperatures that vary between 300 and 450°C [72, 73]. These temperatures are still higher for several plastic substrates, and for this reason a lot of emphasis has been given to methods separating the synthesis process from the final fabrication of the electronic devices. For example, recently transfer and contact printing have been explored to overcome thermal budget issues ([23, 30, 53, 74]. These methods allow synthesis of the NWs using the conventional VLS method on substrates that can withstand high temperatures (e.g., Si, SiC, sapphire) followed by transferring the NWs on flexible substrates, where the rest of the low-temperature processing steps (e.g., metallization by printing or lithography) can be carried out to obtain electronic devices. Detailed discussion of these methods can be found elsewhere [49, 75], including another Cambridge Element focussing on this topic [31].

This Element is organised as follows: Section 2 presents the VLS mechanism for Si NWs with a focus on the fundamentals of growth and phase diagrams, experimental techniques and growth kinetics for Si NWs through atomistic steps. The growth and characterization of CNTs, which is another 1D nanostructure explored extensively for flexible and large-area electronics, is discussed in Section 3. Further, the growth and characterization of ZnO NWs, which is another 1D nanostructure explored extensively for sensing, electronics and energy applications in flexible and large-area electronics, is discussed in Section 4. Finally, the conclusion in Section 5 summarises the key points from this Element.

2 VLS Growth Mechanism for Si NWs

2.1 Fundamentals of VLS NWs Growth Mechanism

2.1.1 The VLS Concept and Working Principle

VLS NW growth is accomplished by injection of atomic flux into catalyst particles under suitable conditions, and the supersaturated catalyst particles precipitate NWs preferentially at the substrate/catalyst interface (Figure 2.1).

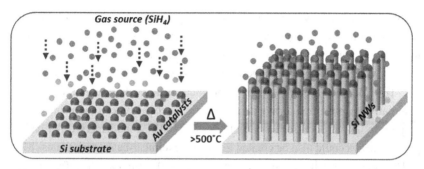

Figure 2.1 Schematic diagram of the catalyst particle–assisted VLS growth mechanism for Si NWs

For example, Au catalyst particles in the range of a few nanometres to several microns in diameter are suitable for Si NW/whisker growth and have been widely studied [76, 77]. In a controlled ambient, Au nanoparticles placed over Si substrates at and above temperatures of 400°C catalyse the NW growth process. Bulk Au metal with 19 at.% Si melts at the eutectic temperature of 363°C, which is explained using a binary Au–Si phase diagram (Figure 2.1). The phase diagrams form the basis for the NW growth using the VLS mechanism. A similar mechanism for CNT growth has been proposed in the case of highly coupled CH_4 plasma to Ni catalysts, causing its surface layers to be rich in carbon, so that the carbon diffuses across the surface of the Ni catalyst to produce CNTs [78]. The experimental conditions suitable for the VLS growth mechanism (Section 2.2) are commonly provided by physical vapour deposition (PVD) tools such as thermal evaporator, e-beam evaporator, pulsed laser deposition, MBE, etc., and chemical vapour deposition (CVD) systems [77]. The vapour flux is injected into the catalyst particles under the controlled ambient (Sections 2.2 and 2.3), and the catalyst particles undergo phase changes over a wide range of temperatures (200 to ~1100°C).

2.1.2 Catalyst Materials and Si NW Growth by the VLS Mechanism

Historically, a wide spectrum of catalyst materials (Pt, Al, Ni, Ti, Fe, Ga, etc.) has been explored for Si NW synthesis [76]. The noble metal nanoparticles (Au, Pt, etc.), however, provide great practical convenience as catalyst particles for Si NW synthesis. Indeed, Au is the most commonly used catalyst for Si NWs due to its chemical inertness during annealing and its easy availability. Thin Au film (~5 nm thick) present over a substrate (e.g., Si, sapphire, steel foil) heated above ~800°C results in statistically distributed Au nanoparticles with an average diameter of ~100 nm. The commonly used Si NW growth conditions

(Table 2.2), which involve a temperature range of 400–600°C, is not ideal to dewet the thin metal films to obtain catalyst particles. Due to the chemical inertness of Au, the nanoparticles could be created under regular inert gases such as Ar, N_2 or forming gas. Dewetting of oxygen-sensitive elements such as Al, Ti, Cu, etc. requires more sophisticated annealing conditions [79]. Alternatively, patterned arrays of nanoparticles could be created using lithographic tools with controlled particle density [80]. Monodispersed Au nanoparticle arrays could also be created using wet chemical-based techniques [81].

The substrates with catalyst nanoparticles ($\sim 10^{10}/cm^2$) are transferred into a NW synthesis tool such as CVD, MBE, laser ablation chamber or e-beam evaporator [82] to obtain the NWs. Note that catalysts such as Au are known to create deep-level defect states and therefore are not allowed in the CMOS process. As a result, it is preferred to remove the Au catalyst from the tip of the NWs after the desired length has been obtained. This is usually achieved with aqua regia–based wet chemical etching. To allow further processing to develop electronics, it is important for the structure of NWs to remain unaffected (i.e., unintentional incorporation) during the removal of metal catalyst nanoparticles [83–85]. During the NW growth process, the catalyst atoms tend to inject into the NW crystal lattice despite poor solubility (see the next section) [86]. In view of the electronic devices, these NW impurities could introduce the trap states and can act as recombination centres for charge carriers and thus deteriorate the device characteristics [87]. Due to this adverse effect, Au is not the best catalyst for NW-based electronics.

Similar to the Au catalyst, Pt nanoparticles could be prepared in a less controlled ambient and transferred into the NW growth chamber. Pt catalyst has the ability to yield NWs from solid and liquid states depending on the temperature of the NW growth process [88, 89]. The experimental conditions could be elucidated using a Pt–Si binary phase diagram. In addition to Pt, several other catalyst materials (Al, Ni, Ti, Fe, Ga, etc.) have been attempted under different NW synthesis conditions to overcome the difficulties posed by Au. Among these, transition metals such as Al, Ni and Ti have been shown to enable the defect-free and kink-free Si NWs needed for device applications. The nanoparticles of these metals must be prepared under oxygen-free conditions and transferred seamlessly into the NW growth chamber such as CVD and MBE. This is because even a thin layer of native oxide (i.e., a few atomic layers) over the surface of these catalyst particles could prevent the growth of NW by screening Si injection from vapour. The plasma etching of native oxide could help prevent this issue in the case of Al as catalyst [77]; however, similar issues arise for other oxide-forming catalysts such as Ti, Ni, Cu and Fe. The injection of catalyst atoms into the NW lattice could also create other positive effects in some cases. For example, unintentional incorporation of Al during the NW growth process could introduce p-type doping in Si NWs [79]. In the case of Al catalyst, the catalyst removal step could be excluded before the device fabrication, as it is beneficial for making metal contacts and, unlike Au, it does not introduce defects.

During the growth process, the metal catalysts need to overcome losses due to evaporation so as to yield straight, nontapered NWs and must possess low vapour pressure in the growth temperature window [90]. In practice, catalyst-assisted NW growth is accomplished through both liquid and solid states of catalyst particles. Metals such as Al, Cu and Pt have been shown to assist NW synthesis using solid [88] and liquid catalysts. The composition of catalyst particles during growth plays a key role in deciding NW growth kinetics. The growth rates with liquid phase catalysts are an order of magnitude higher than with solid catalysts. Liquid catalysts tend to absorb vapour flux efficiently and create a saturated homogeneous solution due to easier movement of atomic flux inside the catalyst solution. This contributes to a faster growth rate in comparison with the solid catalyst particles. Metal catalyst–NW material binary phase diagrams provide insight into catalyst composition, phase and temperature for NW growth. These are described in the next section.

2.1.3 Role of Phase Diagram in the Synthesis of Si NWs

Binary phase diagrams play a crucial role in the selection of catalyst materials and provide a roadmap to experimental conditions for catalyst particle–assisted NW growth. The NW synthesis results from a combination of the catalyst and the NW elements at suitable temperature and pressure conditions given by the phase diagram. During this process, the nanoparticles of catalyst materials undergo different compositional and phase changes. For example, in the VLS process, solid catalyst particles are converted into liquid solution by incorporation of atoms from vapour sources, and these liquid catalyst solutions maintain a dynamic equilibrium between the source vapour and solid NWs to yield a steady growth of NWs [91–93]. The equilibrium composition and the phases dictate the growth kinetics of resulting single crystalline NWs. The phase change occurring in the catalyst particles during the NW growth process could be identified through the equilibrium phase diagrams [94]. There are two types of phase diagrams commonly available in the literature to explain the catalyst particle–assisted NW growth. For the case of Si NW growth, the two types are:

Type 1: NW growth from liquid phase catalyst particle (e.g., Au–Si, Zn–Si and Ga–Si)

Type 2: NW growth from solid and liquid phase catalysts (e.g., Al–Si, Cu–Si, Ti–Si and Pt–Si)

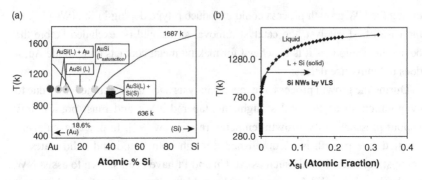

Figure 2.2 (a) Binary Au–Si phase diagram with tie line above the eutectic temperature. (b) Bulk Ga–Si phase diagram with low solubility of Si in Zn. (Reprinted with permission from [76, 95, 96]).

Type 1: NW growth from liquid phase catalyst particle The binary phase diagrams exhibit various phases and compositions with respect to temperature. Figure 2.2(a) shows a gold–silicon binary phase diagram with a tie line along which the VLS Si NW growth process occurs. Au metal catalyst is best suited due to the eutectic feature of the Au–Si binary phase diagram and complete immiscibility in the solid phase [76]. The eutectic point (T_E) in the Au–Si system presents at 363°C and 19 at.% Si (Figure 2.2(a)). The catalyst particles need to be heated over the eutectic temperature for the growth of Si nanowires. The horizontal tie line (Figure 2.2(a)) above the eutectic temperature helps in understanding the phase transformation of Au–Si to nucleate Si NWs. The horizontal line in the LHS of eutectic temperature (363°C) has pure Au solid phase in the starting point. This tie line progresses towards RHS by adding Si atoms to meet the liquidus line where it is eventually converted into a liquid Au–Si solution. This nano Au–Si liquid solution is considered to be a homogenous solution without any concentration gradient. Further, Si addition to this Au–Si solution takes the system to the liquidus boundary on the Si side. This will lead to the Au–Si system's being saturated with Si material. Essentially, the RHS of the tie line with respect to the second liquidus phase boundary consists of Au–Si (liquid) + Si (solid)).

At temperatures above eutectic point, addition of Si to Au–Si solution beyond the liquidus line tends to produce **"pure Si"** solid material. The solubility of Au in Si is shown to be zero in the phase diagram. These predictions keep the binary phase diagrams a central tool in the VLS-nanowire growth research. Figure 2.2(b) shows a binary Ga–Si phase diagram, which has characteristics similar to those of the Au–Si system with different combination of temperature and solubility level [95, 96]. As shown, solubility of Si in Ga is very low alongside the melting point of pure Ga at 43°C. The Ga–Si phase diagram has eutectic temperature at 30°C

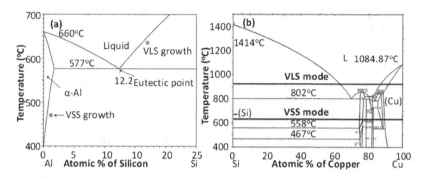

Figure 2.3 Binary phase diagrams (a) Al–Si; (b) Cu–Si. VLS and VSS growth regions are depicted by arrows (a) and tie lines (b). (Reprinted with permission from [76, 98].)

with < 0.01 at.% Si. This proves the lowest solubility of Si for any reported catalyst materials. Ga catalyst provides two important advantages for Si NW growth: (1) NW growth could be carried out at low temperatures (150–200°C); (2) the low solubility of Si is advantageous for creating NW heterostructures with sharp interfaces. Additionally, Ga–Si liquid solution is stable over a wide temperature range, and Ga could serve as a p-type dopant for Si NW devices.

Type 2: NW growth from solid- and liquid-phase catalysts Catalyst choices such as Al [97] and Cu [98] are advantageous for Si NW growth due to their compatibility with the CMOS device fabrication process. Binary phase diagrams of Al and Cu have a eutectic feature with different levels of Si solubility (Figure 2.3(a)–(b)). Both catalysts (Al and Cu) have shown Si NW growth above and below the eutectic tie line. As seen in these phase diagrams, catalyst particles remain in the solid state during growth at sub-eutectic temperature, which is popularly known as the vapour–solid–solid (VSS) mechanism [88]. Al catalyst particle aids Si nanowire growth by VLS and the VSS mode, as indicated in the phase diagram.

The eutectic point (T_E) of Al–Si system is at a temperature of 577°C with 12.6 at.% of Si. A finite solubility of Si in Al always exists. Hence, Si nanowire growth using solid AlSi particles occurs below 577°C by VSS mode, and above this temperature VLS mode produces Si NW similar to the Au–Si system. VLS mode produces kink-free, nontapered Si NW with <111> orientation using conventional CVD conditions. Conductivity studies on these VLS-grown NWs have demonstrated the presence of Al in the Si NWs lattice. This is due to unintentional injection of Al from the catalyst particles to NWs lattice (p-type dopant) during growth. Incidental formation of native oxide over Al surface, however, limits its wide utility as a catalyst. Similarly, the binary Cu–Si phase diagram (Figure 2.3(b)) shows eutectic temperature at 802°C, which has been used to synthesise Si NW by both VLS and VSS modes. As shown in Figure 2.3 (b), the Cu–Si system consists of many intermediate silicide phases below the eutectic temperature. Cu particles undergo a minimum of four phase transitions for the growth of Si NW by VSS mode. The pure Cu particle first transforms into Cu_5Si upon Si addition and is subsequently converted into Cu_3Si. At sub-eutectic temperatures (<802°C), the available solid phases are termed as η-Cu_3Si, η'-Cu_3Si and η''-Cu_3Si, which mediate Si NW growth under CVD growth conditions. Temperatures above 802°C allow Si NW growth through liquid Cu–Si solution by VLS mode. Hence, Cu catalyst material allows the synthesis of Si NWs at a wide temperature range using solid and liquid catalysts.

2.2 The Experimental Techniques for the Growth of Si NWs

Since the first demonstration of 1D Si whiskers by VLS mechanism, reported using vapour-phase chemical technique [63], a number of physical and chemical growth techniques have been developed for the synthesis of NWs – with both the top-down or subtractive and the bottom-up methods. Since the VLS growth mechanism is associated with the latter, the physical and chemical methods related to the bottom-up method are discussed below. A summary of physical and chemical techniques is presented in Table 2.1. Details related to physical and chemical methods related to top-down or subtractive strategies can be found elsewhere [29, 30, 32, 48, 56, 74, 99–102].

2.2.1 Physical Methods for Synthesis of Si NWs

The physical techniques supply atomic flux in the form of either vaporised atoms or sputtered plasma, which serves as building blocks for bottom-up grown NWs. It is necessary to hold good control over supplied atomic flux to produce sub-100 nm diameter NWs for many applications [24, 103, 104].

The key physical techniques for bottom-up synthesis of NWs include traditional thin-film deposition systems such as physical evaporation (thermal and e-beam), sputtering, laser ablation and MBE. Thin-film techniques for the synthesis of NWs use a wide range of vacuum conditions with pressures ranging from 10^{-3} to 10^{-12} Torr. These techniques employ substrate heaters to achieve the required temperature (300–700°C) for growth of most semiconducting NWs. Thin-film techniques use solid source materials, which are vaporised and fed into the catalyst surface as elemental atoms or plasma. These techniques set the requirement of solid source materials such as sputtering targets and evaporation sources. The availability of the compositional controlled "pure"

Table 2.1 Summary of physical and chemical techniques for the synthesis of semiconducting NWs (Si NWs, ZnO NWs and CNTs)

S.No	Technique	Source material	Advantages	Disadvantages
			Physical Techniques	
1	Evaporation (Thermal & e-beam)	Filaments, wires, boats, etc.	Low cost	Unsuitable for compound semiconductors
2	Sputtering	Solid targets	Low cost; availability of wide range of target materials	Typically, unsuitable for oxygen-sensitive materials (Si, Ge or III-V nitrides)
3	Laser ablation	Solid targets	Suitable for sub-20 nm wires	Unsuitable for wide range of materials; difficult to control NW diameter
4	Arc discharge method	Solid targets	Low cost and high yield for μm-long CNTs; can produce SW or MW CNTs	Impurities due to side products; no control over CNT orientation or diameter
5	Molecular beam epitaxy (MBE)	Elemental evaporation sources	Suitable for wide materials	Expensive technique; low growth rates
6	Lithography and dry etch techniques	Wafers	Precisely defined array of NWs	Expensive tools; requirement of wafers and cleanroom for processing

Table 2.1 (cont.)

S.No	Technique	Source material	Advantages	Disadvantages
			Chemical Techniques	
7	CVD, PECVD, PTCVD, LPCVD and APCVD	Vapour molecules and gases	Single-crystal compositional controlled NWs	Harmful gas sources; expensive; vacuum processes
8	Gas phase techniques (for CNTs, ZnO)	Gaseous molecules	Large quantity synthesis	Oxygen-sensitive materials could not be synthesised
9	Chemical beam epitaxy	Gases and vapours	Suitable for wide materials	Expensive method
10	Metal assisted chemical etching (MACE)	Wafers	Low cost and simple to execute	Requirement of wafers

solid source materials limits the applicability of these techniques for NW synthesis.

Implementation of heating sources to produce high temperatures (>500°C) under vacuum conditions is a drawback of physical thin-film techniques. MBE is one of the successful techniques to produce kink-free NWs of semiconductor materials such as Si and GaAs. The growth rate of NWs using MBE is in the range of 1–2 nm/min, which is an order of magnitude lower than the vapour-phase techniques [58]. MBE is a preferred technique for the formation of abrupt NW heterojunctions with great control in growth rates [105]. In addition to the drawback of low growth rates, however, the MBE poses difficulties in terms of demanding maintenance of expensive equipment and high vacuum conditions, etc. These disadvantages keep the technique from being a major tool for synthesis of NWs. The laser ablation is another important physical technique to produce Si NWs with a moderate growth rate of ~1 μm/min [106, 107]. The technique uses a laser source to ablate atoms from a composite target consisting of NW material and catalyst. For example, a solid target with composition of $Si_{0.9}Fe_{0.1}$ ablated using Nd-YAG, 532 nm laser source can produce Fe metal–catalysed Si NWs with diameter 6–20 nm at a temperature of 1200°C [106]. The disadvantages of this technique are the nonavailability of wide target materials, poor control over NW dimensions and location of growth. The CNTs are also synthesised with laser ablation and arc discharge methods, where a solid carbon source is evaporated at high temperatures (3000–4000°C) and CNT growth occurs upon condensation of these vapours. In CVD, a transition metal (Fe, Ni, Co, etc.) catalyst film is annealed in the temperature range of 600–1100°C to form nanoparticles, and a hydrocarbon gas is passed through the furnace, which results in the growth of CNTs on catalyst particles. Thermal evaporation has been used for synthesis of catalyst-free oxide NWs such as ZnO, VO_x and In_2O_3 [108]. In this method, the metal oxide powders are sublimated in the evaporation chambers and allowed to condense as NWs over a substrate that is kept at low temperature. The growth mechanism of V–S-grown NWs is based on the screw dislocation model [109], and dense NWs can be grown over large areas by the vapour–solid growth mechanism without catalyst particles [110]. Thermal evaporation is one of the cost-effective physical techniques but largely limited to synthesis of metal oxide NWs, and control over uniformity of diameter is an issue with this method. Techniques such as e-beam evaporation are known to produce thin films of a few nanometres in thickness. Sputtering techniques (DC and RF) can generally produce stoichiometric-controlled thin films from a few nanometres to tens of microns. This method, however, is not useful for synthesis of semiconducting NWs such as Si, Ge, GaAs and GaN.

Most of the envisaged NW-based applications require single crystalline NWs with well-controlled diameters. From the electronic device point of view, the controlled doping is another important requirement. In light of these requirements, the atomistic process during the early and steady-state growth of NWs carries significant importance.

2.2.2 Chemical Methods for Synthesis of Si NWs

Starting with the synthesis of micron-sized (diameter) Si whiskers in 1964 [63], over the past five decades, the semiconducting NWs of varying composition from elemental to ternary materials have been grown using chemical techniques [58]. The vapour-phase chemical techniques offer several advantages for NW-based applications. The main advantage lies in the ability to supply a wide variety of sources (gases and liquid precursors) of NW materials, with the CVD process allowing accurate control over flux, wide temperature range (200–1200°C) and higher growth rates for commercialization. The gases and vapours used as precursors in chemical vapour techniques could be precisely controlled using modern electronic gas flow systems. The CVD experimental growth conditions for Si NWs and CNTs are summarised in Table 2.2, and various variants of gas-phase techniques are discussed below.

The thin-film deposition process in CVD consists of four basic atomistic steps [111, 112]: (1) mass transport of source gas molecules to the growth front through lamellar flow; (2) decomposition of gas molecules at the substrate surface. Temperature-dependent gas phase decomposition/reaction is prevented to avoid particulate formation; (3) surface diffusion followed by

Table 2.2 Summary of experimental conditions for Si NW and CNT growth by CVD technique

1D Nanostructure	Si NW		CNT
	Exp. Condition 1	Exp. Condition 2	
Catalyst	Au, Cu, Ti, Al	Au, Al, Ti	Fe, Ni, Co
Source	$SiCl_4$	SiH_4	CH_4, C_2H_2, C_2H_4, CO
Carrier Gas	H_2	H_2	H_2, NH_3, N_2, Ar
Growth Temperature	800–1100 (°C)	400–650 (°C)	600–1100 (°C)
Growth Pressure	~100 (Torr)	< 1 (Torr)	1 – 760 (Torr)

attachment of atoms at available lattice sites for crystal growth, and (4) removal of reaction by products/unreacted source gases from the chamber. These four basic atomistic steps of the thin-film growth process could be directly extended for catalyst particle–assisted NW growth. Catalyst particle–assisted VLS and VSS mechanisms are largely executed by the CVD technique using a broad range of experimental conditions. The NW growth is carried out in well-controlled ambient, as most of the CVD precursor sources are harmful (flammable and toxic). For example, single crystalline Si NWs have been synthesised by the VLS mechanism using hydride (SiH_4) and chloride ($SiCl_4$) precursors in the temperature range between 300 and 1100°C (Figure 2.4). The precursor dissociation temperature range and suitable catalyst phase for NW growth must coincide in this process to yield the wires. The NW synthesis processes have been carried out using a wide span of pressures ranging from 100 mTorr to 760 Torr (Table 2.2) using low- and atmospheric-pressure CVD systems.

Typically, the CVD systems have well-controlled chamber pressure consisting of (inert) carrier and source gases, which provides measured amounts of reactive molecules for material growth. The mechanistic

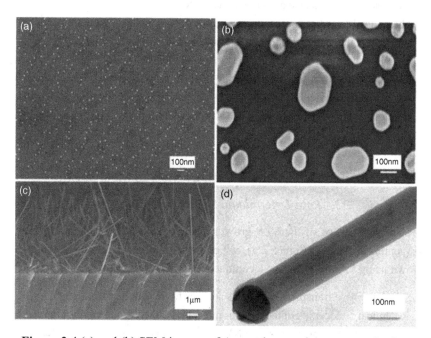

Figure 2.4 (a) and (b) SEM images of Au catalyst particles prepared using dewetting of thin film. (c) Si NWs grown by CVD using a $SiCl_4$ precursor at 900°C. (d) TEM image of the Si NW with Au catalyst at the tip.

insights have been obtained by studying the growth process in situ by feeding gaseous sources through a Transmission Electron Microscopy (TEM) column [83, 113]. The temperature and pressure of the chamber play a crucial role in the CVD-assisted NW growth process. Variants of the CVD system offer conditions suitable for growing of different types of NWs. For example, the hot-wall CVD systems maintain uniform temperature of the reaction chamber for high throughput. Thermal CVDs specialise in conformal deposition of thin films for a wide range of material systems under controlled gaseous ambient and surrounding heat source. Modern commercial CVD systems engineered with innovative chamber geometries allow precise control of gases and reaction zone. Gases can be introduced into the chamber of these modern CVDs in either the horizontal or the vertical configuration with controlled substrate rotation. The cold-wall CVD systems control localised heating at the substrate zone. NW synthesis in the UHV microscopic column for in situ analysis is one of the classical examples of cold-wall systems [114, 115]. Summarising, the technical advantages of the CVD technique for NW synthesis outweigh other tools in the following aspects:

- It is a well-established commercial technique with diverse liquid and gaseous precursors to synthesise a wide variety of NWs.
- The control over the source gas flux allows NW synthesis with atomic-level control on the composition and allows growth of novel axial NW hetero-structures with sharp material interfaces [19, 116, 117].
- The possibility of in situ doping during the NW growth adds a huge advantage for NW-based devices. The gas phase injection of dopants can be carried out simultaneously with the growth process to obtain uniform distribution of dopants [118–120].

2.3 Kinetics of Si NW Growth by the VLS Mechanism

2.3.1 Rate-Determining Step of the VLS Process

The atomistic steps of the VLS growth mechanism can be divided into four sequential processes (Figure 2.5), which are common for any NW material grown using the vapour-phase technique [121, 122]:

1. Vapour-phase transport of gaseous sources to the catalyst front.
2. Chemical dissociation of the gaseous molecules at the vapour–catalyst interface followed by the injection of the NW feedstock into the catalyst particle. This step has an associated energy barrier for gaseous molecules to get incorporated into the catalyst particle.

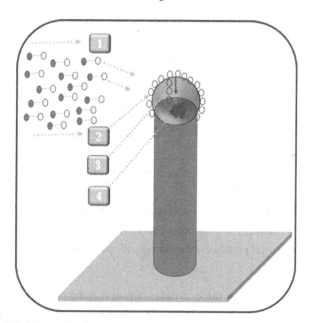

Figure 2.5 Schematic diagram of the four basic steps in VLS NW growth process: (1) vapour transport, (2) atoms adsorption and incorporation at the V–L interface, (3) diffusion of atomic flux inside the catalyst particle, (4) crystallization at the L–S interface.

3. Diffusion of atoms inside the catalyst particle to reach at catalyst–NW interface. The calculated timescale for this process is in picoseconds for a catalyst size in the range of 10–100 nm. For kinetic analysis, a homogenous nanoliquid catalyst solution is considered during NW growth.

4. Crystallization of NW at the catalyst–NW interface, through various modes. The growth modes are dependent on the saturation level of the catalyst particle.

The above four basic steps could be further divided into the many injection and ejection processes that will be detailed in the next section. The important aspect of this basic division is to identify the critical steps that control VLS NW growth. Step 1 is neglected in the literature, as it is not an activated process that can act as a bottleneck of the VLS process. Similarly, step 3 is an atomic diffusion process that occurs in a picosecond timescale in a nanosized catalyst particle, which is unlikely to control the growth process. The literature reports have argued that either step 2 (dissociation and injection [123]) or step 4 (crystallization at the interface) could form the rate-determining steps. This

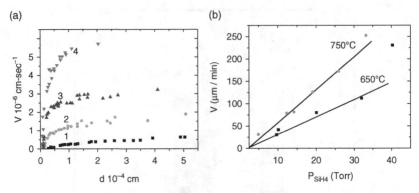

Figure 2.6 (a) Dependence of growth rate over NW diameter. Plots (1–4) show increase in gas phase (SiCl$_4$) supersaturation. (b) Growth rate dependency over the precursor partial pressure. (Reprinted with permission from [121, 123].)

debate has continued for three decades through rigorous experimental and theoretical works. In this regard, Giviargizov's [121] 1975 report is a benchmark for physical chemistry modelling of VLS growth process (Figure 2.6(a)). The experimental work shows that the growth rate of <211> oriented whiskers is higher than the <111> direction whiskers under similar growth conditions. The growth kinetics work reported for Si and GaAs NW grown under atmospheric CVD conditions has led to the conclusion that the crystallization (step 4) is the rate-controlling step. The crystallization step as a bottleneck for VLS process is counterargued by many reports such as the popular one led by Bootsma and Gassen through experimental studies [123]. They have observed that the growth velocity is linearly dependent on the precursor partial pressures (Figure 2.6(b)). Similar observations are made by other research groups, with the conclusion that the precursor incorporation at the vapour–catalyst interface is the rate-determining step of the VLS growth process [124]. Recently, this debate has taken a new direction with application of a steady-state kinetics approach to the VLS growth mechanism through a phenomenological model by Schmidt *et al.* [91, 125]. The VLS growth steady state is explained by a multistep chemical reaction X→Y→Z, where the final product Z forms through an intermediate compound Y from the initial reactant X [125]. At steady state, the rate of formation of Y is equal to the rate of formation of Z. Similarly, the VLS mechanism is a multistep process in which a balanced Si injection–ejection results in the NW growth. At steady state, the rate of injection flux into the catalyst is equal to the rate of ejection. For Si NW growth by the VLS mechanism, this implies that at steady state the interplay between Si incorporation and crystallization results in a steady Si concentration gradient in the catalyst droplets. The NW growth is driven by this saturated Si concentration in catalyst droplets. The model explains that the injection and the ejection of atoms through the catalyst depend on the concentration of solute atoms in catalyst particles. This phenomenological hypothesis took the idea of a single rate-determining step to a new direction. Equal rates of the steady-state injection–ejection process rule out the possibility of one step being much slower to control the growth process. During steady-state growth, the atomic flux of NW material goes through different energy levels from vapour phase to solid state. The chemical potential of the catalyst particles lies between vapour flux and NWs. The detailed chemical potential analysis is presented in the next section.

The growth rate of NWs is dependent on their diameter when it is in the range of 5–150 nm (Figure 2.6(a)) and the growth rate tends to decrease with diameter of the NWs. Many works have observed this trend with conventional CVD conditions [121, 126, 127], which indicates that the NW growth velocity has an initial non-steady transient state followed by a steady-state regime. The steady-state regime allows the NW lengths to be a linear function of time. The saturated growth velocity is proportional to the gas phase saturation in the system as depicted in Figure 2.6(a). The radius dependence can be explained by defining the effective supersaturation as follows:

$$\Delta\mu = \nabla\mu_{vs} - \frac{2\Omega\gamma^{vs}}{R} \tag{2.1}$$

where $\Delta\mu_{vs}$ is the chemical potential difference between vapour and bulk solid with infinite radius ($R = \infty$), Ω is the atomic volume and γ^{vs} is the interfacial energy of the vapour–solid interface. The second term contributes to the reduction of chemical potential due to growth of finite-radius (R) NWs, which is commonly known as the Gibbs–Thomson effect. The second term in Eq. (2.1) can dominate at smaller radius (i.e., R < 20 nm) and therefore under these conditions the net effect will be a reduction of $\Delta\mu$ and hence of the growth rate of the NWs. The radius dependence is attributed to the Gibbs–Thomson effect. This widely accepted theoretical foundation, however, cannot explain the following developments in kinetics of NW growth:

a) Experimental observations on NW growth under UHV conditions have shown that the growth rate of NWs is independent of diameters in the range of 10–150 nm [124]. This observation could not be explained using radius-dependent supersaturation.

b) Traditionally, the definition of supersaturation, $\Delta\mu$, assumed that the vapour phase and the catalyst liquid solution are in equilibrium during the steady growth of NWs. Theoretical studies show, however, the existence of chemical potential difference between the atomic flux in the vapour and liquid solutions [91]. This means that supersaturation driving NW growth needs to be redefined.

c) The growth process is driven by the supersaturated nanosized catalyst solution to cylindrical NWs. The model of gas-phase supersaturation-driven wire growth should be used to predict the growth rates.

These observations led to the formulation of detailed thermodynamic analysis of the VLS growth process, which has helped explain convincingly the contrary observations regarding the diameter-dependent growth rates of NWs.

2.3.2 Thermodynamics of the VLS Growth Process: Au-Catalysed Si NW Growth by the VLS Mechanism

During synthesis, the Si atoms go through various energy levels from vapour phase through Au–Si liquid and finally they are incorporated into solid NWs. The energy cascade during the process is shown in Figure 2.7 with bulk Si as reference [92, 93, 128]. R represents the radius of the hemispherical Au–Si droplets and Si NWs. A major benefit of this energy diagram is that it defines the thermodynamic driving force for the VLS NW growth process. The driving force and supersaturation help to estimate the NW growth rates theoretically by the VLS mechanism. As described in the earlier sections on phase diagrams, the NWs crystallise from a supersaturated Au–Si liquid solution of equal dimension.

Based on the energy cascade depicted in Figure 2.7, the effective supersaturation with catalyst–vapour equilibrium is expressed as follows:

$$\Delta\mu_{vs} = kT \ln\left(\frac{P}{P^0}\right) - \frac{2\Omega}{R}\gamma^{Si} \tag{2.2}$$

where $\Delta\mu_{VS}$ is the effective supersaturation or chemical potential difference of Si between vapour and NW, k is Boltzmann's constant, T is the temperature, P is the partial pressure of Si in the vapour, P^0 is the equilibrium vapour pressure of Si, Ω is the atomic volume and γ^{Si} is the Si–gas interfacial energy. The effective supersaturation and hence the growth rate is dependent on the radius of the NWs, which is apparent from the second term of Eq. (2.2). The phenomenological kinetic model indicates that the chemical potential of Si in the vapour

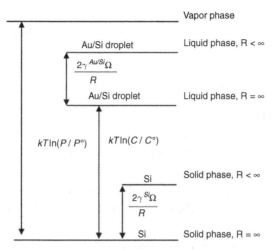

Figure 2.7 Energy cascade of the Si during the NW growth process by the VLS mechanism. (Reprinted with permission from [128].)

phase is higher than in the finite-sized Au–Si droplet. Hence, the steady-state supersaturation is expressed from Eq. (2.2) as:

$$\Delta\mu_{LS} = kT \ln\left(\frac{C}{C^0}\right) + \frac{2\Omega}{R}\left(\gamma^{\frac{Au}{Si}} - \gamma^{Si}\right) \tag{2.3}$$

where $\Delta\mu_{LS}$ is the effective supersaturation or chemical potential difference of Si between Au–Si droplets and Si NWs, C is the concentration (atoms/m^3) of Si in the Au droplet, C^0 is the equilibrium concentration and $\gamma^{Au/Si}$ is the liquid droplet–gas phase interfacial energy. As the typical values of C/C^0 fall in the range of 1.1 to 1.3, the first term in Eq. (2.3) can be of the order of 10^{-21} J/atom or higher. The value of Ω/R in the second term in Eq. (2.3) is in the order of 10^{-21} m^2, which becomes significant when the radius of the NW is less than 10 nm. The surface energies of Au–Si nanodroplets and NWs are roughly about the same value (1–2.3 J/m^2). These calculations show that the second term is insignificant, as its value is two orders lower than that of the first term. Hence, the effective supersaturation driving the NW growth is given by:

$$\Delta\mu_{LS} = kT \ln\left(\frac{C}{C^0}\right) \tag{2.4}$$

The dimensional effects are cancelled as the growth process occurs between two nanostructured entities (Au–Si catalyst/NWs). The diameter dependencies under specified conditions could arise from other sources such as nucleation frequency. Different modes of nucleation at interfaces are elaborated in the next section.

2.3.3 Atomistic Processes in the Kinetics of NW Growth

The steady-state NW growth is governed by various injections and ejections of atomic flux through the catalyst particles (Figure 2.8). The details of atomistic processes for liquid catalyst particle–assisted Si NW growth, with SiC_4 and SiH_4 precursors, are presented in this section. The processes are equally valid for the growth of other NW material systems, however, such as CNTs, ZnO, Ge, GaAs, GaN, etc. [129–134]. The importance of each Si injection and ejection step is quantitatively described here. The steps related to Si injection are as follows: (a) vapour transportation to the catalyst front; (b) chemical dissociation of $SiCl_4$ or SiH_4 at the catalyst–vapour interface followed by injection of Si into the nanodroplet (catalytic adsorption); (c) adsorption of source molecules at the NW sidewalls or substrate followed by the surface diffusion for Si injection (surface diffusion); and (d) diffusion of Si atoms inside the catalyst particles to reach the interface. The steps related to Si ejection from the catalyst are as follows: (e) evaporation of Si from the catalyst particle (evaporation); (f) loss of Si due to reverse reaction with gas phase under CVD growth conditions (reverse reaction); (g) Si crystallization at the interface (crystal growth) (Figure 2.8).

During steady state, the rate of Si injection will be equal to the rate of ejection [91, 128, 135]. A balanced injection and ejection of atoms through the Au catalyst will result in a steady-state concentration to drive the NW growth process. The steady state can be expressed as follows:

$$\left[\frac{dC}{dt}\right]_{Injection} = \left[\frac{dC}{dt}\right]_{ejection} \tag{2.5}$$

where C is the concentration of atoms expressed in number per unit volume in the catalyst droplet and t is the time. With all the listed components included, Eq. (2.5) could be expanded as

$$\left[\frac{dC}{dt}\right]_{Adsorption} + \left[\frac{dC}{dt}\right]_{Surface\ diffusion}$$
$$= \left[\frac{dC}{dt}\right]_{Evaporation} + \left[\frac{dC}{dt}\right]_{Reverse\ reaction} + \left[\frac{dC}{dt}\right]_{Crystal\ growth} \tag{2.6}$$

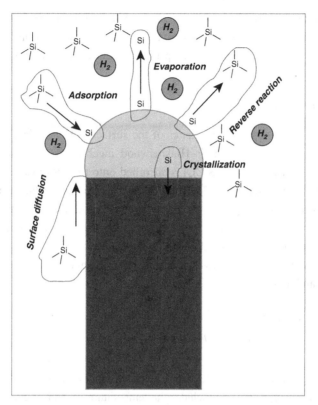

Figure 2.8 Illustration of the detailed description of all the atomistic steps involved in the VLS NW growth mechanism

The vapour transport (i.e., step 1 of Si injection) and Si diffusion (i.e., step 4 of Si injection) are neglected in kinetics studies for the reasons discussed in the previous section. The catalyst particle is assumed to be a homogeneous liquid solution with no concentration gradient existing within the nanovolume. This condition is practically attained in systems such as Au–Si, Au–Ge, Al–Si, etc. when they are above their eutectic temperatures. Alternatively, the steady state must be applied to the 2D interface [136] (catalyst /NWs) when NWs are grown using solid catalysts by using the vapour–solid–solid (VSS) growth mechanism. The detailed expressions for the injection–ejection steps will be explained in the following section.

2.3.3a Injection

Catalytic adsorption In the NW growth process, vapour flux ($SiCl_4$ or SiH_4) constantly flows through the catalyst surface with a certain partial pressure maintained inside the system [137]. The flow dynamics results in a localised partial pressure in the vicinity of the catalyst (i.e., Au–Si solution) particles. The gaseous species are physically adsorbed over the surface of the catalyst and dissociate chemically to inject Si into the catalyst particle. The reaction pathway is expressed as follows:

$$SiH_4(g) \rightarrow Si(Au) + 2H_2(g)$$

The adsorption–desorption kinetics of the surface-enhanced reaction could be studied using the Langmuir–Hinshelwood mechanism. A monolayer of gaseous molecules (SiH_4 or $SiCl_4$) is adsorbed onto the sites available on the surface of the catalyst particle. The rough surface of the liquid catalyst is ideal as it offers maximum adsorption sites compared to the solid catalyst particles. This is the key reason behind the higher growths of NWs obtained using a liquid catalyst as compared to a solid one. The concentration of Si adsorbed in the catalytic decomposition process causes a dynamic change in the Au–Si droplet, and this can be expressed as follows:

$$\left[\frac{dC}{dt}\right] = \left[\frac{K_1 P}{1 + K_2 P}\right]_{droplet} \frac{A_{droplet}}{V_{droplet}} \tag{2.7}$$

where P (typically, $P \ll 1$) is the partial pressure of the source gases, K_1 (m/J⁻ˢ) is the reaction rate constant and includes the temperature-dependent energy barrier associated with dissociative adsorption (Eq. (2.7)) [138], and $A_{droplet}/V_{droplet}$ is the surface-to-volume ratio of the catalyst droplet. The value of K_1 for the silane decomposition [139] is calculated using the following expression:

$$K_1(SiH_4) = 2622 \exp\left(\frac{-Q_R}{R_G T}\right) \frac{1}{kT} \tag{2.7a}$$

where Q_R is the energy barrier for decomposition of SiH_4 over the Au–Si droplet. The catalytic decomposition data is extrapolated from the epitaxial growth of the Si films using a SiH_4 precursor.

Surface diffusion The second source of atoms for growth is through gaseous decomposition at the NW sidewalls followed by surface diffusion. This mechanism gained importance for the growth of compound semiconductor NWs (e.g., GaAs, GaN, ZnO, etc.) where two components have to be incorporated for growth. The source molecules are decomposed over the NWs' sidewalls within a diffusion length of λ, and this flux tends to inject into the catalyst particle. The concentration changes in catalyst particles due to surface diffusion flux can be expressed as follows:

$$\left[\frac{dC}{dt}\right] = [K_1 P] \frac{2\pi R\lambda}{V_{droplet}} \tag{2.8}$$

The gaseous molecules absorbed beyond the diffusion length tend to desorb and mix with the gas phase. The diffusion length could be calculated using the following expression:

$$\lambda = a\exp\left(\frac{E_{desorption} - E_{Surface\ diffusion}}{2kT}\right) \tag{2.8a}$$

where E represents the activation energy barrier for desorption and surface diffusion processes and a is the average interspacing between two adsorption sites (i.e., interatomic separation). The calculated values of λ fall between 22 and 110 nm for the temperature range of 900–1200 K, which is the range used for growing Si NWs with SiH_4 and $SiCl_4$ precursors. Arguments based on past experimental observations, however, suggest that the surface diffusion component plays a minor role in the injection of atomic flux into the catalyst droplet [140, 141]. This VLS mechanism fails to sustain when there is no catalyst particle available for NW growth. Given these facts, the two possible scenarios where adsorption followed by surface diffusion could play a role are (1) growth of NWs of less than 100 nm in diameter, and (2) growth of binary NWs where the second component is added by surface diffusion.

2.3.3b Ejection

Evaporation The loss of Si atoms from the Au–Si liquid solution likewise is a common phenomenon which is also confirmed by in situ microscopy studies [83]. The evaporation causes catalyst particles to shrink during NW growth and could lead to tapering of NWs. Evaporation of Si from the hemispherical catalyst droplet is expressed as [142]:

$$\left[\frac{dC}{dt}\right] = \left(\frac{\Omega C P_{eq}^0}{(2\pi mkT)^{1/2}}\right)\exp\left(\frac{2\gamma^{lv}\Omega}{RkT}\right)\frac{A_{droplet}}{V_{droplet}} \tag{2.9}$$

where P^0_{eq} is the vapour pressure of Si in equilibrium with bulk Si melt, ΩC is the Si content in the nano Au–Si solution and the exponent term is due to the Gibbs–Thomson effect. The estimated evaporation rate of Si from the Au–Si droplet at 900°C is less than 1 atom/s. This rate is insignificant for NWs having diameter less than 10 nm and when growth is performed under UHV conditions.

Atom ejection due to reverse reaction The atomic species in the catalyst particle reacts with the adsorbed molecules eventually ejected into the gaseous phase. For example, Si atoms in the Au–Si droplet react with the adsorbed Cl or H to form $SiCl_4$ or SiH_4, typically at temperatures above 900°C. This is due to the chemical potential difference between vapour and catalyst. The reaction pathway could be expressed as follows:

SiH_4 (g) or (adsorbed) $\Leftrightarrow$ Si (g) or (adsorbed) $+ 2H2$ (g) or (adsorbed)

and

Si (g) or (adsorbed) $\Leftrightarrow$ Si (catalyst)

In general, the injection–ejection of Si from Au–Si droplets depends on the Si concentration in the droplet. The chemical composition of Si in the droplet is difficult to measure due to experimental difficulties. Hence, the expressions for the reverse reaction ejection are deduced by assuming thermodynamic equilibrium between the catalyst droplets and the vapour phase, i.e., $\Delta\mu_{vl} = 0$. The rate of reverse reaction is expressed as follows:

$$\left(\frac{dC}{dt}\right) = (K_1)\frac{A_{droplet}}{V_{droplet}}\left(\frac{C/C^0}{P/P^0}\right)^2\left(\exp\left(\frac{2\gamma^{Au/Si}\Omega}{RkT}\right)\right)^2 P^S \tag{2.10}$$

As depicted, the ratio between the supersaturation of gas and condensed phases decides the ejection by reverse reaction, where P^S is the partial pressure of Si in close proximity of the catalyst droplet.

Crystal growth of NWs Atom ejection followed by crystallization is central to the VLS process. The atomistic process of nanocrystal growth occurs through different modes and is governed by respective expressions [143]. The rate of change in Si concentration within the catalyst particle due to NW growth is expressed as follows:

$$\left(\frac{dC}{dt}\right) = \frac{1}{V_{droplet}}\frac{\pi R^2}{\Omega}\frac{dh}{dt} \tag{2.11}$$

where dh/dt is the growth rate of NWs obtained by various modes (Figure 2.9). The nucleation and growth of NWs commonly proceed through a layer-by-layer (LL) mechanism (Figure 2.9(a)). For example, 2D crystals of Si nucleate at the interface and spread to complete a layer before the next layer precipitates from

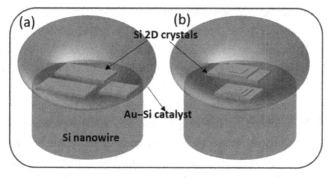

Figure 2.9 Schematic illustration of NW growth modes at the catalyst–NW interface: (a) Layer-by-layer mode. (b) Multilayer mode.

the Au–Si droplets. There are two possibilities for this process to occur in a system: (1) Nucleation occurs at and around the middle of the interface (IF); (2) Nucleation initiates at the edges of the interface and spreads towards the centre of the catalyst–NW or substrate interface. This mode is described in the NW literature as nucleation and growth through a triple-phase boundary (TPB) [144].

Each layer in Figure 2.9 is considered to be one atomic layer thick. The growth rate of the NWs by LL mode is expressed as follows:

$$\left(\frac{dh}{dt}\right)_{LL-IF} = \left(J\pi R^2\right)a \tag{2.12}$$

$$\left(\frac{dh}{dt}\right)_{LL-TPB} = (J2\pi R)a \tag{2.13}$$

where R is the radius of NWs or the catalyst–substrate interface, J is the nucleation rate for unit area (IF) or unit length (TPB) and a is the height of the nucleated layer. Eq. (2.12) and (2.13) could be used to calculate the growth rate of NWs under a given experimental condition. The nucleation rate is dependent on many parameters, and this is expressed as follows:

$$J_{IF} = \left\{N_0\nu\Omega C^0\exp\left(-\frac{Q_D}{kT}\right)\right\}\frac{C}{C^0}\left(\ln\left(\frac{C}{C^0}\right)\right)^{1/2}\exp\left(-\frac{\pi(\chi a/kT)^2}{\ln(C/C^0)}\right) \tag{2.13a}$$

where N_0 is the number of atomic sites (10^{19} m^{-2}) available for attachment, v is the vibrational frequency (10^{13} s^{-1}) of the atoms and χ is the edge energy (10^{-10} Jm^{-1}) of the step. These parameters are constant for Si nucleation from Au–Si liquid solution. Q_D is called desolation energy, i.e., the barrier for Si atoms to detach from Au–Si melt to attach to Si crystals. The value of Q_D is observed to vary between 5 and 20 kT. Another influencing parameter over J is the supersaturation, C/C°, which could be estimated by balancing the injection–ejection process. This data is deduced using incubation studies on NW nucleation from the Au–Si solution over a heterogeneous sapphire substrate [113, 145, 146]. The LL mode of nucleation and growth is experimentally verified for NW growth using a solid catalyst in the VSS growth mechanism [136].

The LL mode of growth occurs either as a single nucleation event followed by lateral propagation or as multiple nuclei joined together to complete a layer (Figure 2.9(a)). It is clear that the small NWs ($\ll 100$ nm) could complete a layer with fewer nucleation points at the interface. As the catalyst diameter increases above hundreds of nanometres, the nucleation mode tends to change from LL to multilayer mode (ML) (Figure 2.9(b)). In ML mode, a new layer nucleates before completion of the previous layer (Figure 2.9(b)). Alternatively, higher supersaturation also promotes growth by ML modes.

2.3.4 Estimation of Growth Rate of NWs Using Kinetic Model Parameters

The atomistic mechanisms provide insights into the key parameters that control NW growth rate. The measured and numerical estimates for key parameters allow theoretical estimation of NW growth under given experimental conditions. These parameters are summarised below.

Catalyst size The influence of the catalyst particle radius over the growth rate is widely studied under different experimental synthesis conditions. At smaller NW diameters (<10 nm), the curvature-dependent Gibbs–Thomson effect acts so as to reduce the growth rate of NWs. Also, the catalyst size decides the 2D interfacial (catalyst–substrate) area available for nucleation to occur.

Supersaturation (C/C°) It is evident from Eq. (2.12) and (2.13) that NW growth rate is highly dependent on the supersaturation condition in the catalyst prevailing during the growth. The measurement of Si concentration in the Au–Si catalyst solution during NW growth is difficult with nanocharacterization tools [145]. The C/C° values typically vary from 1.1 to 1.8, and the range could be estimated using binary phase diagrams. Supersaturation tends to decrease with increase in diameter of the NW, as observed in the growth rate saturation at

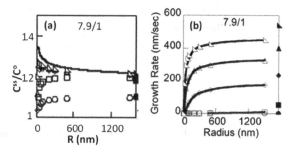

Figure 2.10 (a) Estimation of Si supersaturation (C/C°) and (b) NW growth rates predicted for different catalyst radius using Eq. (2.12). The kinetic parameters (Q_R/Q_D) are depicted on the top. (Reprinted with permission from [128].)

larger diameters (~500 nm). The expressions for the derived atomistic mechanisms (Eq. (2.7)–(2.13)) are balanced to extract supersaturation (Figure 2.10(a)) for an experimental condition.

Kinetic parameters: **(a) Energy barrier for adsorption** The injection of atoms at the vapour–catalyst interface needs to overcome an energy barrier, which is denoted as Q_R (Eq. (2.7a)). The Q_R value should be used for the catalytic decomposition barrier during NW growth, and this information is absent in the NW growth literature. The closely associated data is that found for the heterogeneous decomposition of $SiCl_4$ or SiH_4 for the growth of epitaxial Si films. The estimated value of Q_R for $SiCl_4$ decomposition is 79 kcal/mole. This value helps to approximately evaluate the upper bound value for desorption barrier for Si gases/vapours over liquid Au–Si droplets.

(b) Desolvation energy barrier The nucleation of 2D layers from the supersaturated liquid solution requires overcoming the desolvation energy barrier, Q_D. The values of Q_D reported for nucleation from bulk melts fall in the range of 1–10 kT. The NW growth over sapphire substrates in the early stage of growth helped to estimate Q_D values for Si nucleation from Au–Si solutions. The calculated values are in the range of 5–15 kT. By using these kinetic parameters, the growth rates of NWs have been estimated (Figure 2.10(b)) to be within an order of magnitude with respect to data reported in the literature.

The major aim of the kinetic analysis is to elucidate the dependencies of NW growth rate. The balance between injection–ejection processes will result in a steady-state concentration of Si in the Au–Si droplet. The estimation of Si supersaturation can be obtained by equating the expressions explained in the previous sections, and the steady state can be expressed as Eq. (2.5):

$$\left[\frac{dC}{dt}\right]_{Injection} = \left[\frac{dC}{dt}\right]_{ejection}$$

By substituting Eq. (2.7), (2.10) and (2.11), the expression becomes:

$$\left[\frac{K_1 P}{1 + K_2}\right]_{droplet} \frac{A_{droplet}}{V_{droplet}} = \left[(K_1) \frac{A_{droplet}}{V_{droplet}} \left(\frac{C/C^0}{P/P^0}\right)^2 \left(\exp\left(\frac{2\gamma^{Au/Si}\Omega}{RkT}\right)^2 P^S\right)\right]$$

$$= \frac{1}{V_{droplet}} \frac{\pi R^2 dh}{\Omega \, dt} \tag{2.14}$$

The value of $C/C^°$ is calculated for the range of catalyst radii from 5 to 1500 nm. The range covers the diameters which are significant for devices as well as synthesis of large whiskers. The experimental conditions are used from Si NW growth using a $SiCl_4$ precursor above 900°C. Figure 2.10(a) shows the variation of $C/C^°$ with respect to NW diameter for various values of gas-phase supersaturation. The values of $C/C^°$ are deduced by equating expression (2.14) using a MATLAB code for the given set of kinetic parameters and gas-phase supersaturation. The estimated NW growth rate using Eq. (2.12) and (2.14) is observed to fall within an order of magnitude (1–1000 nm/s). The trend in the growth with respect to the catalyst size explains the contrary observations in the literature. Importantly, the supersaturation prevailing in the droplet is shown to be a strong function of catalyst particle size. The kinetics studies clearly demonstrate the factors (atomistic and macroscopic) influencing the growth rate of NWs. This will help to design a synthesis process to obtain NWs with precise control over diameters.

3 Carbon Nanotubes

3.1 Structure and Properties

Subsequent to the discovery of fullerenes [148], growth of carbon nanotubes (CNTs) using the arc discharge method was reported in 1991 [149]. Since then, CNTs have attracted much research interest due to their unique properties such as high current-carrying capacity, high aspect ratio, ballistic conduction, high mechanical strength, high thermal conductivity, etc. [150–152]. CNTs can be viewed as rolled-up sheets of graphene forming cylindrical structures comprising either single-walled CNTs (SWCNTs) or multiwalled CNTs (MWCNTs), as

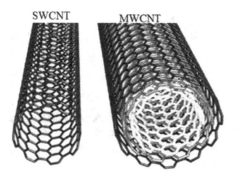

Figure 3.1 SWCNTs are a single sheet of graphene rolled up to form a tubular structure, whereas MWCNTs are made of several concentric graphitic shells. (Reprinted with permission from [147].)

shown in Figure 3.1. SWCNTs contain a single shell which can be metallic or semiconducting, whereas MWCNTs consist of several concentric shells separated by the interlayer distance of graphite (0.34 nm). The diameter of SWCNTs generally lies between 1 and 2 nm, whereas that of MWCNTs lies between 2 and 30 nm, and both come in bundles with lengths up to millimetres. Graphene and CNT are sp^2 hybridised, and they have many common characteristics. The chirality, curvature and resulting quantum confinement along the circumference of the tube, however, provide CNTs some unique properties, which are different to those of graphene [152] [153]. The 'chirality' of the tube is defined by the chiral vector **C** along the circumferential direction of the tube and the chiral angle θ made by **C** with the zigzag direction, as shown in Figure 3.2(a)–(b) [154]. **C** can be written as the sum of the unit vectors $\mathbf{a}_1$ and $\mathbf{a}_2$, defined along the two zigzag directions of the hexagon [152]:

$$\mathbf{C} = n\mathbf{a}_1 + m\mathbf{a}_2 \qquad (3.1)$$

where (n, m) are the number of steps along the zigzag direction of the hexagonal graphene lattice. The angle θ determines the amount of chirality or helicity in the tube. When $\theta = 0°$, $(n, 0)$, the tubes are known as zigzag CNTs, and when $\theta = 30°$ (n, n), they are called armchair CNTs due to the formation of carbon atoms around the circumference of the tube as illustrated in Figure 3.2(b). For all other angles between $0°$ and $30°$, $(n \neq m \neq 0)$, nanotubes are known as chiral CNTs. The electronic band structure of a graphene sheet near the Fermi energy is given by an occupied π band and an unoccupied π^* band. These bands meet at the Fermi level at K points in the Brillouin zone (Figure 3.3(a)), rendering graphene as a semimetallic or zero-gap semiconductor. On the other hand, CNTs can be

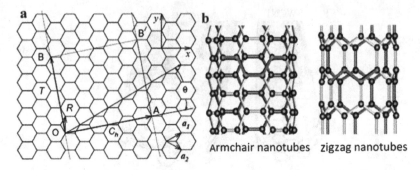

Armchair nanotubes   zigzag nanotubes

Figure 3.2 (a) Honeycomb graphene sheet which gives a CNT structure upon rolling up. The chirality of the tube is determined by the chiral vector C_h and the chiral angle θ made by C_h with the zigzag direction. (Adopted with permission from [155]. (b) Schematic of armchair and zigzag structures. Reprinted with permission from [154].)

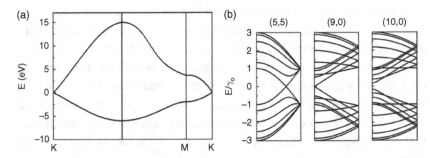

Figure 3.3 (a) The electronic band structure of graphene calculated for π and π^* bonds in the tight binding model. The conduction and valence bands meet at K points in the Brillouin zone. (b) Band structures for (5, 5) and (9, 0) nanotubes have zero band gaps, which indicate the metallic nature of the tubes, whereas a (10, 0) tube shows a finite band gap, indicating a semiconducting tube. $\gamma_o = 2.9\,eV$. (Reprinted with permission from [157].)

metallic or semiconducting, depending on their chirality. The condition for a CNT to be metallic is:

$$n - m = 3d \tag{3.2}$$

where d is an integer. This means that, if $n - m$ is a multiple of 3, the CNT is metallic, and otherwise it is semiconducting. Following this rule, all armchair CNTs (n, n) are metallic whereas all zigzag $(n, 0)$ and chiral tubes can be metallic or semiconducting. For a random production batch of SWCNTs, two-thirds are expected to be semiconducting [156] [152].

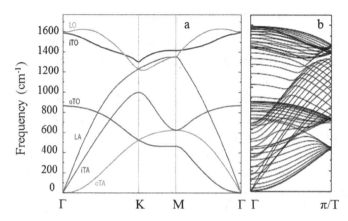

Figure 3.4 (a) Phonon dispersion relation of graphene. (b) Phonon dispersion relation of a (10, 10) armchair SWCNT. (Reprinted with permission from [155, 158].)

Band structures of an armchair (5, 5) tube and two zigzag tubes, (9, 0) and (10, 0), are shown in Figure 3.3(b). For the (5, 5) and (9, 0) tube, the valence band and the conduction band have zero gap, so the tubes will be metallic in nature. On the other hand, (10, 0) tube shows an energy gap, so the tube will be semiconducting. In case of CNTs where $d \neq 0$, the degenerate point moves a little away from the K point due to the effect of large curvature of the tube and results in a small gap. Armchair tubes (n, n) are always metallic as they are independent of curvature due to their symmetry. With increasing radius R of the nanotubes, the band gaps of semiconducting and small-gap tubes decreases as 1/R and $1/R^2$, respectively [152] [156].The two carbon atoms in the unit cell of graphene results in six phonon branches which contain three acoustic phonon and three optical phonon modes as shown in Figure 3.4(a) [158]. The sp^2-bonded carbon materials are characterised using these phonon dispersion relations. The iTO and LO optical modes are degenerate at the zone centre, Γ, and belongs to the Raman active E2 g mode [156]. For CNTs, the zone folding scheme is applied to get the phonon dispersion relations. The unit cell of a SWCNT contains 2N carbon atoms and there are 6N degrees of vibrational freedom. In Figure 3.4(b), the phonon dispersion relation for a (10, 10) armchair tube is shown. For a (10, 10) tube, N = 20, so 120 vibrational degrees of freedom are expected; however, because of the mode degeneracies, only 66 phonon branches appear [156].

3.2 Electrical Conduction in CNTs

The molecular dimensions along the circumferential direction, long tubular structure and chirality-dependent electronic behaviour of CNTs incorporate a variety of unique transport properties in them, which make CNTs suitable for various electronic devices [159]. CNTs provide an excellent platform to study the fundamental properties of a one-dimensional quantum wire where the electron–electron interaction plays a significant role in the transport properties. By now, many characteristics of CNTs as 1D conductors have been observed experimentally, which include room-temperature ballistic transport [151], coulomb-blockade [160], Aharanov-Bohm effect [161], quantised conduction [162, 163], etc. The resistance r of a conventional conductor with cross section A and length L is given as $r = \rho \frac{L}{A}$, where ρ is the resistivity of the material. The ohmic behaviour remains valid until the size of the material remains much larger than each of the following characteristic lengths: (1) the de Broglie wavelength, which depends on the kinetic energy of electrons, (2) electron mean free path, which is the average distance travelled by an electron before it scatters and (3) the phase relaxation length, which is the distance that an electron travels before its initial phase is lost. When the material dimensions become comparable with one of these characteristic lengths, however, the resistivity becomes length independent because of the quantum confinement effects, and the electrons act like waves, which show interference effects depending on the boundary conditions, defects and impurities in the material [156].

When the electron mean free path, l_{mfp}, becomes larger than the length of a wire, the conduction in the wire will be ballistic. The wire in this case can be regarded as an electron waveguide whereby each mode (conduction channel) contributes an exact amount G_o in the total conductance of the wire. G_o, called 'quantum conductance', is independent of material properties and is given as follows [156]:

$$G_0 = \frac{2e^2}{h} = 77.48 \times 10^{-6} \text{ S} \tag{3.3}$$

where e is the electronic charge (1.6×10^{-19} C) and h is Planck's constant (6.63×10^{-34} Js). The inverse of G_0 is the quantum resistance R_o and is given as:

$$R_o = \frac{h}{2e^2} = 12.906 \text{k}\Omega \tag{3.4}$$

A SWCNT is considered as a room-temperature 1D ballistic conductor. Metallic CNTs often show this quantised conductance, whereas only a proportion of semiconducting CNT devices have come up within this theoretical limit at room temperature [163–165]. As there are two conduction channels in a SWCNT due to the band degeneracy, the total quantum conductance of a tube having l_{mfp} greater than its length will be $2G_o$, and its resistance will be $R_o/2$ [156]. If the length L_{cnt} of a CNT is larger than l_{mfp}, however, then its total resistance will scale with its length as follows [166]:

$$R_t = R_c + \frac{R_0}{2}\left(\frac{L_{cnt}}{l_{mfp}}\right) \approx R_Q \frac{L_{cnt}}{l_{mfp}} \tag{3.5}$$

where R_c is the additional contact resistance that arises due to the poor contact and can be neglected for perfect contacts. $R_Q = R_0/2 = 6.45$ kΩ is the quantum resistance of a SWCNT. The mean free path of electrons in carbon nanotubes depends on scattering rate caused by the structural defects (impurities) and phonons. The electron–phonon scattering depends on the temperature and on the applied bias whereas the electron–impurity scattering depends on the defect density (quality of CNTs), and its range has been reported from 100 nm to 100 μm in the CVD-grown CNTs [167].

3.3 Synthesis of Carbon Nanotubes

Since the discovery of CNTs produced by the arc discharge method in 1991, several other CNT growth techniques have been developed, which can be categorised under three major techniques: (1) arc discharge method, (2) laser ablation method and (3) chemical vapour deposition (CVD). In the arc discharge and laser ablation methods, a solid carbon source is evaporated at high temperatures (3000–4000°C) and synthesis of CNTs occurs upon condensation of these vapours. In CVD, a catalyst film is heated, usually in the temperature range of 600–1100°C, and a hydrocarbon gas is introduced in the furnace, which results in the growth of CNTs on catalyst particles. A brief overview of these techniques is given below.

3.3.1 Arc Discharge Method

The first growth and identification of CNTs were made by utilising this technique [149]. Evaporation of carbon rods in an Ar-filled vessel (100 Torr) using DC (direct current) arc discharge resulted in the production of MWCNTs of lengths up to 1 μm and diameter ranging from 4–30 nm (Figure 3.5(a)–(b)) at the negative electrode [149]. The schematic of the arc-discharge apparatus is shown in Figure 3.5(c) [168]. In this technique two carbon rods separated by 1–2 mm are placed end to end in a chamber, usually filled with a gas (Ar, He, CH$_4$, H$_2$) at pressures between 50 to 700 mbar [169]. When DC current (50–100 A) is passed between

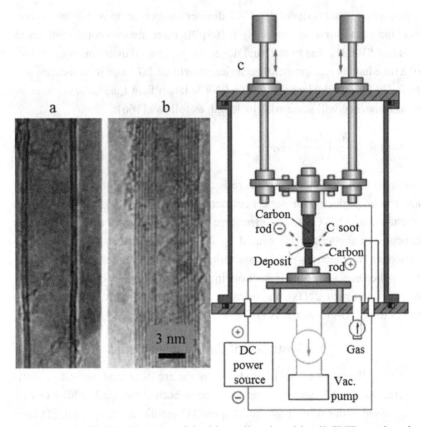

Figure 3.5 (a, b) TEM Images of double-wall and multiwall CNTs produced using arc discharge of carbon rods by Ijima in 1991. (Reproduced with permission from [149].) (c) Schematic of arc-discharge apparatus for the growth of CNTs. (Reprinted with permission from [168, 175].)

the electrodes, high-velocity electrons emitted by the cathode bombard the anode, which increase the anode's temperature and subsequently evaporate the source. Part (20–40%) of the evaporated carbon is converted to MWCNTs upon condensation on the cathode. The arc-discharge-produced CNTs are usually in the form of bundles due to van der Waals interactions, are of good crystalline quality and have lengths in microns and diameters of 4–30 nm. Using this method, large-scale synthesis of MWCNTs has been achieved [170, 171]. A drawback of this technique is the production of large quantities of by-products like fullerenes and amorphous carbon which not only lower the yield but also require purification of MWCNTs. The purification is usually done by heating the as-grown material in the presence of oxygen to oxidise the by-products [172]; however, it also removes a considerable amount of CNTs and damages the quality of the tubes.

Research is being carried out to produce pure carbon nanotubes by adopting different methods. About 70% pure MWCNTs has been achieved using the arc-discharge method in liquid nitrogen [71]. High-purity synthesis of MWCNTs up to 95% has been achieved by using magnetic fields around the arc plasma to control the synthesis [173, 174]. SWCNTs can also be grown using metal catalysts such as Fe, Co, Ni, etc. [173]. But a major disadvantage is that the grown nanotubes contain a lot of metal particles and require extra treatment to purify.

Researchers are trying to improve the synthesis and parameter control of SWCNTs by optimising the chamber pressure, controlling the anode-to-cathode distance and by using different catalysts [176].

3.3.2 Laser Ablation Method

The energy density of a laser pulse is much higher than any other vaporising technique and so is suitable to vaporise materials having high boiling points such as carbon. Smalley's group developed the laser furnace method, together with an annealing system for the growth of large quantities of fullerene and CNTs in 1992 [177, 178]. In this technique a pulsed or continuous laser is used to vaporise the graphite target at high temperature in an oven. The main difference between a pulsed laser and a continuous laser is that a pulsed laser requires very high laser intensity (around 100 kW/cm^2) as compared to the continuous laser (around 12 kW/cm^2). A typical laser ablation set-up consists of a laser furnace, a quartz tube with a window, a catalyst-doped carbon target, a water-cooled system and a pressure control system to maintain a constant flow rate of buffer gas, usually at 500 Torr, as shown in Figure 3.6 [168]. A laser beam (YAG or CO_2 laser) is focused by passing through the window of a quartz tube onto the target to vaporise it in a buffer gas (Ar/He) environment. Upon condensation, carbon atoms and molecules tend to form large clusters. The catalyst particles, however, which have a relatively slow condensation rate, attach to these clusters and prevent their closing into a cage structure [70]. The produced CNTs are accumulated in a trap due to the gas flow and collected. MWCNTs are synthesised if pure graphite is used, whereas a catalyst (Co, Ni, Fe) is mixed with graphite for the synthesis of SWCNTs. The laser ablation method has the ability to produce high-quality and purer (up to 90%) SWCNTs with controlled diameters [168]. SWCNTs with high crystallinity, minimal defects and low contamination have been achieved by utilising the laser ablation method and subsequent purification processes [179, 180]. The yield, however, is not as good as that of the arc-discharge method. Research is being done to improve this technique in different ways. Ultrafast pulses from the free electron

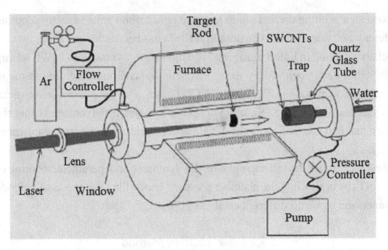

Figure 3.6 Schematic of laser ablation technique used to produce CNTs. (Reprinted with permission from [168].)

laser (FEL) method and the continuous wave laser-powder method are examples of the developments in laser ablation techniques [181, 182].

3.3.3 Chemical Vapour Deposition

The chemical vapour deposition (CVD) is widely used in the semiconductor industry to produce high-quality thin films of different materials such as silicon, tungsten, carbides, oxides, nitrides, etc. After the discovery of CNTs in 1991 by the arc discharge method, CVD has been largely used for the growth of CNTs by utilising its different technical variants [173, 183, 184]. Synthesis of CNTs by CVD involves pyrolysis of a carbonaceous substance in the presence of a metal catalyst such as Fe, Ni, Co, etc. under controlled conditions in a furnace, as shown in Figure 3.7. The CVD growth of CNTs can be viewed in two steps. The first step involves deposition of the catalyst on a substrate followed by the activation of catalyst particles by chemical etching, thermal annealing or plasma treatment. The size of the catalyst particles, density and the material used for the catalyst affect the growth profile of carbon nanotubes. In the second step a carbonaceous substance is vaporised to the reaction chamber for the actual growth of CNTs. The temperature required for the growth of CNTs in CVD usually ranges from 600 to 1100°C [185–188]. The choice of carbon feedstock, carrier gas, flow rate, chamber pressure, substrate, catalyst material, catalyst particle size, catalyst density, catalyst support material and temperature are the main parameters which affect growth rate, quality, density, diameter and selectivity (SWCNTs or MWCNTs) of carbon nanotubes.

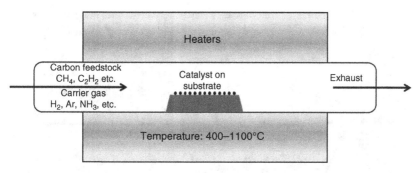

Figure 3.7 Schematic of CVD setup where a carbonaceous gas mixed with carrier gases is introduced at high temperature (600–1100°C) into the furnace and growth of CNTs occurs on a catalyst-deposited substrate.

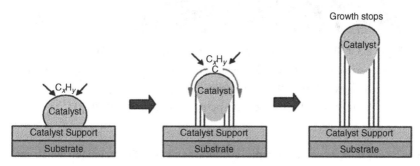

Figure 3.8 Schematic of a growth mechanism of carbon nanotubes: decomposition of carbon feedstock, diffusion and graphitization; steady growth of CNT; growth stops. (Reprinted with permission from [190].)

A commonly referred CNT growth mechanism is illustrated in Figure 3.8 and is usually described in three steps: (1) decomposition of hydrocarbons on the metal catalyst surface to form metastable C_2 molecules, (2) diffusion of carbon atoms in the catalyst and (3) precipitation of solid carbon atoms at the other side of the catalyst for the growth of carbon nanotubes [173, 184, 189, 190]. If the catalyst particles are not strongly attached to the surface of the substrate, they can be lifted up with the growth of nanotubes, which is referred to as the "tip growth" mechanism, whereas the growth of CNTs in which metal particles remain attached to the surface of the substrate is known as a "base growth" or "root growth" mechanism [184]. It is assumed in one explanation that the metal particles, on which deposition takes place on one half of the surface, are spherical or pear-shaped [191]. Generally, arc-discharge- and laser-ablation-produced CNTs are of higher crystalline quality as compared with CVD-grown

Figure 3.9 SEM images of patterned and vertically aligned CNTs using a photothermal CVD (PTCVD) system at 2 Torr pressure of C_2H_2/H_2. (a, b) Patterned growth of CNTs. (c) Vertically aligned CNTs produced on Si substrate. (Reprinted with permission from [199–201].)

CNTs. CVD, however, is a simple, economic and tuneable technique which can be scaled easily because of its high yield and low cost-production capabilities. A key advantage of using CVD is that it has the potential to grow vertically aligned and patterned CNTs at predefined substrate locations, which is not possible with the laser ablation and arc discharge methods. Figure 3.9 shows SEM images of vertically aligned and patterned carbon nanotubes grown on Si substrate. Over the years, remarkable progress has been achieved in the development of cost-effective and high-yield scalable CNT growth processes. With the advent of the 'super growth' process, growth rates higher than 200 μm/min have been achieved, which can produce 2-mm-long vertically aligned CNT mats in just 10 minutes of growth duration [153, 192, 193]. CNTs with lengths around 20 cm also have been reported [194]. Several CVD growth processes have been scaled up for the commercial production of CNTs with a production capacity as high as 16 kg/h [184]. Synthesis of MWCNTs as well as SWCNTs has been achieved using CVD techniques [195–199].

Various techniques have been developed for the growth of carbon nanotubes such as thermal CVD, plasma-enhanced CVD (PECVD), photothermal CVD (PTCVD) [73, 199, 200, 202], aerogel-supported CVD [203], laser-assisted CVD [204], alcohol catalytic CVD [205], remote plasma CVD, etc. [202]. In the

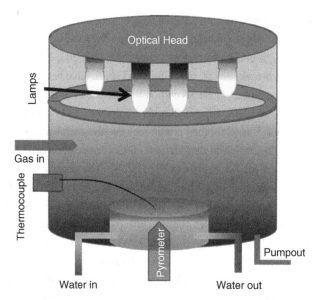

Figure 3.10 Schematic of the photothermal chemical vapour deposition (PTCVD) system. (Reprinted with permission from [200].)

thermal CVD technique, heaters (hot filaments) are used to chemically decompose carbonaceous gas for the growth of CNTs, usually at elevated temperatures (700–1000°C). The plasma-enhanced CVD (PECVD) technique was originally introduced in the semiconductor industry, as it has the advantage of allowing deposition at low temperatures (room temperature to 300°C). In PECVD, a glow discharge is produced to dissociate precursor gases in a reaction furnace by applying a high voltage across the two electrodes. Apart from the ability of PECVD to dissociate the precursor gases, CNT growth does not occur below 550°C. There is enough evidence, however, that PECVD-grown CNTs are well aligned vertically, probably due to the presence of an electric field [206–208]. In PECVD, inert gas (Ar) plasma or H_2 plasma is generally used for catalyst pretreatment before admitting the carbonaceous gas into the reaction chamber.

Many processes in semiconductor device fabrication require temperatures below 550°C, as higher temperatures may cause unwanted reactions and gas-phase decomposition, which subsequently can degrade device functionality. Therefore, to develop in situ CNT-based electronics, high-quality carbon nanotubes are needed to grow at CMOS-compatible temperatures. The NEC group of University of Surrey along with Surrey Nanosystems Ltd. have built a unique state-of-the-art system termed the photothermal CVD (PTCVD) system (Figure 3.10) in which the substrate temperature can be controlled below 400°C. In PTCVD, a water-cooled substrate holder and a thermal barrier layer of Ti/Cu/TiN coated on a Si substrate is utilised to keep the bulk temperature of the substrate low, while an array of lamps in the optical head delivers energy from the top onto the catalyst for the growth of high-quality carbon nanotubes [72, 199, 202, 209]. All physical forms of carbon feedstock, that is, gas, liquid and solid, have been utilised for the growth of carbon nanotubes. Gaseous carbon feedstock can be introduced directly into the CVD chamber [210], whereas liquids are first heated in a flask to vaporise them and a carrier gas is used for their admittance into the chamber [183, 211, 212].

Volatile solids like camphor, naphthalene, ferrocene, etc. directly turn to gas and have been used for CNT growth [213, 214]. The most commonly used precursors for the growth of carbon nanotubes are C_2H_2, CH_4, CO, C_2H_4, benzene, camphor and ethanol [184]. The choice of carbon feedstock affects CNT growth profile [214]; gases require low pyrolysis energy (C_2H_2, C_2H_4 etc.), generally resulting in the growth of MWCNTs, whereas the growth of SWCNTs can be achieved using only selected gases like CO, CH_4, etc., which require high pyrolysis energy [184]. The general explanation for this difference is believed to be the high curvature of relatively small diameters of SWCNTs, which in turn requires high formation energy to overcome the strain energy of carbon atoms. Similarly, the use of alcohol as a carbon feedstock results in almost amorphous carbon-free growth of CNTs where OH radicals are believed to play a role in the removal of excess carbon during the growth process [211]. Moreover, linear hydrocarbons (CH_4, C_2H_2, etc.) generally result in the growth of straight, hollow CNT structures, whereas cyclic hydrocarbons (benzene, xylene, etc.) produce relatively curved and internally bridged CNTs [184].

The most commonly used catalyst materials for the growth of carbon nano-tubes are Fe, Ni and Co [184, 189]. The catalytic ability of these materials for CNT growth is believed to be due to the high diffusion and solubility of carbon in them. The nanoparticles of these materials have a strong affinity towards carbon atoms of growing CNTs, which is believed to be the reason for the formation of the high-curvature tubular structure of CNTs, and that is why CNT diameter is generally found to be equal to the diameter of catalyst nanoparticles [184]. Furthermore, the high melting point and low equilibrium vapour pressure of these materials provide a wide temperature window for the study of carbon nanotube growth. Apart from the commonly used catalyst materials (Fe, Ni, Co), other materials such as Cu, Ag, Au, Pt and Pd also have been used as catalysts for CNT growth [215]. The catalyst material is coated first on a substrate and pretreated prior to the actual growth of carbon nanotubes.

The deposition of a thin layer (usually 0.5–10 nm) of catalyst is carried out either by physical vapour deposition (PVD) techniques such as eva-poration, sputtering, etc. or by solution-based methods. PVD is a commonly used method which involves the vaporization of catalyst material through thermal or electrical energy in a high vacuum or an inert gas environment and subsequent condensation of vapours onto the substrate to form a thin film. The size of the nanoparticles formed by the PVD-deposited catalyst film can be controlled by adjusting the thickness of the film. The use of sophisticated equipment and requirement of high vacuum make this technique relatively expensive. Solution-based techniques involve direct coating of catalyst-containing solution onto the surface of the substrate. The coating is generally performed using one of various techniques such as spin coating, spray coating, dip coating, drop casting, etc. Alcohol-based solutions of metal salts are commonly used because of the high volatility of alcohol [189]. Metal nitrides and acetates generally are used because of their high solubility in alcohol. The drawbacks of solution-based techniques are the requirement of long solution preparation times and the tendency of solution to accumulate at various sites on the substrate [189].

Catalyst pretreatment is performed for the formation of active catalyst nanoparticles, prior to the growth of CNTs. The conditions, however, are not well-defined in terms of temperature and time. In general, catalyst pretreatment time and temperature range from 10 to 60 minutes and 550 to 900°C respectively [210, 216–218]. Although prolonged catalyst pretreatment conditions have been shown to affect CNT diameter and density [219], reduced catalyst pretreatment time for efficient production of CNTs is a requirement for their commercial use and development. Upon annealing, the catalyst film coarsens where catalyst particles agglomerate to form nanosized clusters to minimise their surface energies. The size and the geometry of the clusters depend mainly on the annealing temperature, duration [220], ambient [219] and the catalyst film thickness [210]. The other important aspect is the reduction (deoxidation) of the catalyst particles. This is accomplished, generally, by annealing the catalyst in a reducing gas environment, such as H_2, NH_3, etc. [210, 219]. Hofmann *et al.* studied the CNT growth on a Ta/Fe structure and inferred that in this structure there is no need to use a reducing gas, as Ta acts as a solid-state reducing agent for the formation of active Fe catalyst [221]. Lee *et al.* observed the formation of Al_2O_3 after annealing an Al/Fe bilayer structure in a He gas environment at 700°C, indicating that Al also acts as a reducing agent for Fe [218].

Carbon nanotube growth in CVD is also sensitive to the substrate material. Commonly used substrates are Si, Al_2O_3, quartz, silicon carbide, zeolite, etc. [184, 189]. The most suitable substrate is one that controls the agglomeration of catalyst particles, inhibits catalyst diffusion and prevents catalyst alloying at high temperatures. Alumina has been reported to be a good substrate for CNT growth because of its interaction with the catalyst, which results in uniform dispersion and subsequently high density of CNTs [222]. Si is the most commonly used substrate; however, it hinders CNT growth because of formation of metal (catalyst) silicides and silicates at high temperatures if the catalyst is coated directly over it. A buffer layer, generally SiO_2, is used for the growth of CNTs on Si [189, 223]. Growth of CNTs on metallic layers is relatively challenging, and there are only a few reports of CNT growth on metallic layers as compared to insulating substrates [199, 200, 224]. Commonly used metallic layers for the growth of carbon nanotubes are Cu/TaN/TiN [225], Cu/Ta [147], CuAl alloy [226], TiN [227], Cr (or Ta) [228], etc.

3.4 Controlling Diameter and Chirality of CNTs

Electrical properties of CNTs are highly sensitive to diameter and chirality, so CNT application in electronics requires precise control over these parameters. Selective growth of metallic (m-SWCNTs) or semiconducting (s-SWCNTs) CNTs with controlled diameter, chirality, alignment and density is highly challenging. Also, a convenient method of efficiently characterising chiralities and diameters of individual CNTs in a bundle is not available. Currently, resonant Raman spectroscopy based on Katura plot, TEM and STM is used to evaluate CNT diameter and chirality. Raman spectroscopy is limited by the resonance window and environmental effects, whereas both TEM and STM are inefficient because of their operating conditions and the requirements of sample preparation. There are two approaches to selectively achieving s-SWCNTs or m-SWCNTs; direct selective CVD growth and post-growth separation [229]. Almost all CVD growth parameters, such as temperature, pressure, gas flow, catalyst pretreatment, etc., affect the diameter and chirality of the tubes.

The size of catalyst nanoparticles relates to the diameter of CNTs, and the initial cap formation around the nanoparticle relates to the final chirality of the tubes [200, 230]. Narrow diameter distribution of catalyst nanoparticles reduces CNT diameter and chirality ranges [200]. Ahmad *et al.* demonstrated enrichment of metallic or semiconducting CNTs depending on their growth conditions, as determined by RBM Raman spectra shown in Figure 3.11(a)–(c) [200]. The catalyst nanoparticle size distribution range was narrowed down by adjusting the thickness of the TiN support layer, as shown in Figure 3.11(d)–(g), where 100 nm TiN support layer reduced the distribution range as compared with that of 50 nm TiN. Harutyunyan *et al.* controlled the morphology and coarsening behavior of the catalyst by pre-annealing it in different gaseous environments to enrich the metallic CNT fraction up to 91% [230]. Precise control over catalyst size and its uniform distribution, however, is still highly challenging. Although various methods of selective metallic or semiconducting CNT growth have been reported, so far 100% selectivity has not been achieved.

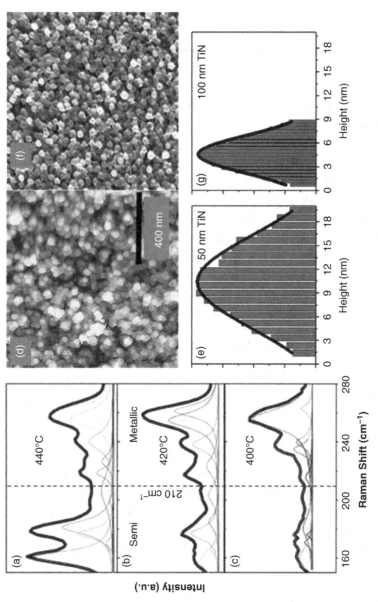

Figure 3.11 (a–c) RBM spectra of the CNTs grown using Fe catalyst on 100 nm TiN support layer to enhance the selectivity of the CNTs at various temperatures. (d, e) AFM image and corresponding Fe nanoparticle distribution on 50 nm TiN support layer. (f, g) AFM image and corresponding Fe nanoparticle distribution on 100 nm TiN support layer. (Reprinted with permission from [200].)

Post-growth separation is another route of achieving the selectivity, through either the non-chemical or the chemical approach. The non-chemical approach includes techniques such as ac-dielectrophoresis [231, 232] and selective breakdown [233, 234].

The chemical approach is the most common and involves CNT functionalisation with chemicals that either form covalent bonds with the CNT π-network or adsorption of surfactant molecules on the walls of CNTs. The separation process involves functionalisation and dispersion in a solution followed by a sorting process. The sorting is based on amplification of the functionalised SWCNT buoyant density and/or physical affinity differences. For example, DNA strands will preferentially wrap around SWCNTs of a certain electronic type or diameter, creating SWCNTs with differential buoyant densities [235, 236], while surfactants such as SDS (sodium dodecyl sulphate) have been shown to generally form ordered micelles with m-SWCNT but not with s-SWCNT, creating SWCNTs that have different physical affinity power [237–240]. Furthermore, by combining certain co-surfactants such as SDS and sodium cholate, SWCNTs can be discriminated based on their diameter [241]. The sorting process that follows is essentially a means to spatially move or segregate the SWCNTs based on their buoyant densities via density gradient ultracentrifugation or based on their physical affinity power via gel agarose electrophoresis and gel agarose column chromatography [237, 241, 242].

4 ZnO Nanomaterials

4.1 Crystal Structure

Under ambient conditions ZnO has a hexagonal wurtzite structure, and its lattice parameters are $a = 0.3296$ nm and $c = 0.52065$ nm [243]. It is usually thought of as successive layers made of O^{2-} and Zn^{2+} ions organised in a tetrahedral order in an alternative arrangement along the c-axis (Figure 4.1) [244, 245]. Despite the neutrality of the entire unit cell of ZnO, some facets can be terminated with positive or negative charges. These facets, called polar facets, have unique growth habits as well as some interesting properties. A common example of ZnO polar facets is the zinc-terminated (0001) facet with positive charge and the oxygen-terminated (000$\bar{1}$) facet with negative charge [243–245].

4.2 Hydrothermal Synthesis of ZnO Nanostructures

ZnO nanostructures can be produced via different methods, which can be categorised into two classes based on their growth temperature. The first one is the high-temperature growth methods, where ZnO nanostructures are grown

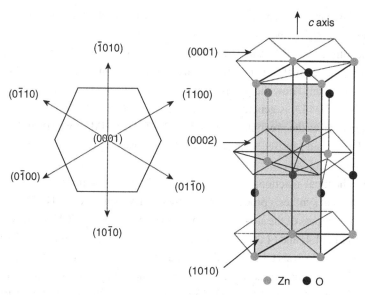

Figure 4.1 Wurtzite structure model of ZnO. (Reprinted with permission from [245].)

at temperatures between 500 and 1500°C. These growth methods are usually performed in a gaseous environment in closed chambers [243]. Examples of the high temperature growth methods are physical vapour deposition [246], metal organic chemical vapour deposition (MOCVD) [247], molecular beam epitaxy (MBE) [248] and pulsed laser deposition (PLD) [249]. The second class of growth methods is the low-temperature methods, where the growth process is performed at relatively low temperatures in the range of 34 to 170°C and usually in a solution environment. Low-temperature growth methods include the hydro-thermal method [250–252] and electrochemical deposition [253]. High-temperature synthesis methods are clearly not the best choice for large-scale production because of many disadvantages such as high energy consumption, high cost to provide the needed growth temperature and the incorporation of catalysts and impurities into the ZnO nanostructures. Consequently, they cannot be integrated with flexible and lightweight substrates [243]. On the other hand, the low-temperature synthesis methods of ZnO nanostructures are a far better option. They hold many advantages including simplicity, lower costs, scalabil-ity for large area growth [254], low processing temperature, non-hazardous nature, catalyst free and the ability to be integrated with the available advanced silicon technologies [243, 255]. Furthermore, the morphologies and properties of the grown nanostructures can be controlled by means of tuning different process parameters [256].The most-used growth solution for hydrothermal synthesis of ZnO nanostructures involves a mixture of a zinc precursor such as zinc nitrate and HMTA in deionised (DI) water. Despite its importance in the hydrothermal growth process of ZnO nanostructures and the large number of reports on it, until now the precise role of HMTA has not been well understood [243, 246], HMTA is generally accepted as a provider of hydroxyl ions in the solution to drive the precipitation reaction [257]. Also, HMTA was thought of as a weak base and a buffer because as the pH increases, hydrolysis rate decreases and vice versa [246].

The chemical reactions shown in Eq. (4.1) to (4.5) summarise the whole hydrothermal synthesis process assisted by HMTA to produce ZnO NWs [257–259]. The outputs of the slow HMTA hydrolysis in water are HCHO and NH_3 as given in Eq. (4.4) and (4.5). It is crucial for the growth process of ZnO NWs that the HMTA hydrolysis take place slowly; otherwise, a significant number of OH^- ions will be produced and subsequently consume the Zn^{2+} ions in the solution due to the increased pH value [257–259]. The formation of zinc hydroxide in the growth solution requires a certain pH level that can be reached by production of NH_3. This resultant of decomposition of HMTA is also important because of its role in stabilising the Zn^{2+} ions in the solution by coordinating with them as shown in Eq. (4.4) and (4.5) [259]. Finally, ZnO is produced through a dehydration process resulting from heating zinc hydroxide reaction (2.8).

$$(CH_2)_6N_4 + 6H_2O \leftrightarrow 6HCHO + 4NH_3 \tag{4.1}$$

$$NH_3 + H_2O \leftrightarrow NH_3H_2O \tag{4.2}$$

$$NH_3 + H_2O \leftrightarrow NH_4^+ + OH^- \tag{4.3}$$

$$2OH^- + Zn^{2+} \rightarrow Zn(OH)_2 \tag{4.4}$$

$$Zn(OH)_2 \rightarrow ZnO(s) + H_2O \tag{4.5}$$

The above reactions are in equilibrium and are controlled by the growth solution concentration, reaction period, temperature and pH level [259]. Generally, the growth solution concentration controls the density and diameters of the NWs. Controlling the reaction period and temperature may lead to morphological changes and affect the aspect ratio of the products. For example, Vayssieres *et al.* [252] in their report on the controlled synthesis of ZnO nanorods found correlations between width and the concentration of growth solution, and between length and growth time. They showed that the length of the nanorods can be increased by two orders of magnitude by simply extending the growth period. With a growth rate of ~1 μm/h at a growth temperature of 95°

C, the maximum length achieved was 10 μm. Another study on the dependence of the length and average diameter of ZnO NWs on the growth time by Greene *et al.* [260] revealed that these parameters are almost linearly dependent on the growth time.

Growing of ZnO NWs on formerly prepared ZnO seeds or structures is much easier than growing them in a solution without seeds or even a substrate. In fact, the possibility of controlling the position of the grown ZnO NWs by positioning the seeds has been reported [261]. The reason behind this is that heterogeneous nucleation is activated at a lower energy level than that of the homogeneous nucleation [252, 259–261]. Therefore, seed layer–assisted heterogeneous growth is much easier than in the case of homogeneous solution and starts at lower levels of supersaturation [252, 259–261]. Seeds or seed layers were also reported to play a major role in controlling the density, diameter and orientation of the NWs [262, 263]. Liu *et al.* studied the influence of the thickness of the seed layer on the morphology of hydrothermally grown ZnO NWs [262]. They were able to control the density of the grown NWs over a wide range (6 orders of magnitude) by simply controlling the seed layer thickness as shown in Figure 4.2. They found that for growth to occur the thickness of the seed layer must be over 1.5 nm. When the thickness of the seed layer was increased from ~1.5 to ~3.5 nm, the number of ZnO NWs was improved by 6 orders of magnitude, and no change in the density was observed beyond this thickness value. Changes in the length and diameter of the NWs were also observed. Furthermore, the authors investigated the effects of these morphological changes caused by controlling the seed layer thickness on the field emission properties of the NWs to reveal the optimum condition. Ma *et al.* reported a controlled hydrothermal synthesis of ZnO NWs by means of controlling the spin. They proved that the density can be improved by 4 orders of magnitude and diameter by 2 orders of magnitude. Seed layers can control the density, length, diameter, orientation and position of the growth. A seed layer, however, is not always good to have in the device structure. In fact, there are many disadvantages of seed layers such as the need of annealing at relatively higher temperature, which limits the ability to use flexible substrates. Also, it prevents direct contact between the grown NWs and the substrate or electrode beneath them, which is important for many device structures like piezoelectric nanogenerators and electron field emitters [243].

Seedless hydrothermal synthesis of ZnO NWs where the growth process starts by nucleating heterogeneously and continues to grow subsequently were reported using materials that possess a small interfacial lattice mismatch with ZnO [264, 265]. Well-aligned ZnO NWs were also reported to grow on GaN that has only 1.8% lattice mismatch with ZnO [264]. Seedless hydrothermal synthesis of ZnO NWs on Si substrates was reported to be

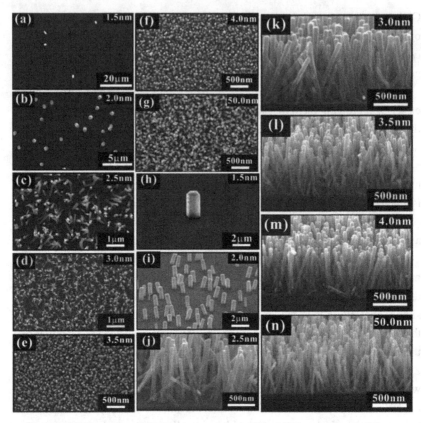

Figure 4.2 Top view SEM images of the grown NWs using a seed layer thickness of (a) 1.5, (b) 2, (c) 2.5, (d) 3, (e) 3.5, (f) 4 and (g) 50 nm. The corresponding tilted and cross-sectional SEM images (h) 1.5, (i) 2, (j) 2.5, (k) 3, (l) 3.5, (m) 4 and (n) 50 nm. (Reprinted with permission from [262].)

difficult owing to the huge lattice mismatch, which is around 40% [243]. One of the popular techniques reported in the literature to control the morphological properties of grown ZnO nanostructures is using shaping or capping materials [243]. Some of these materials assist the axial growth of the crystals by covering all side facets including polyethylenimine (PEI) (Figure 4.3(a)) [266–268] and ethylenediamine [269]. Other capping materials encourage lateral growth by covering the top and bottom surfaces of the crystal. These include Cl^- [270] and $C_3H_5O(COO)_3^{3-}$ (citrate ions) [271]. Xu *et al.* reported a seed layer–assisted hydrothermal synthesis to produce ZnO NW arrays with ultralong NWs for applications in dye-sensitised solar cells [267]. They included PEI as a capping agent in the growth solution and were able to produce the thickest hydrothermally grown

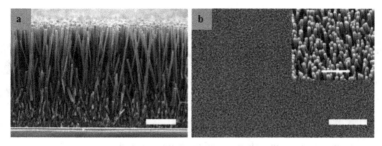

Figure 4.3 (a) Cross-sectional view and (b) top view SEM image of ZnO array. Inset of (b) is a higher magnification top view SEM image. (Scale bar is 5 μm.) (Reprinted with permission from [267].)

ZnO NW array, ~33 μm (Figure 4.3). In comparison with the growth system without PEI, the growth rate was enhanced significantly (more than 10 times) by including PEI. This enhancement in the growth rate was attributed to the role of PEI in directing the growth along the *c*-axis by capping the nonpolar sides of the NWs. Furthermore, the growth solution was observed to be very clear even after the growth process. Authors ascribed this to the fact that PEI may play a role other than extending the length of NWs, which is preventing any homogenous growth in the solution. They claimed that PEI can encapsulate any particle with size less than 20 nm.

They were focusing on the length of NWs with fixed diameter size; however, they did not describe the conditions needed for PEI to adsorb on the nonpolar sides. In addition, the procedure that was followed to remove PEI residuals from the NWs after the growth process involved high-temperature treatment, which is again a disadvantage and has an impact on the properties of the grown products. ZnO nanostructures grown in hydrothermal synthesis methods tend to form 1D structures, since the crystal growth is faster along the *c*-axis than other directions. By selective adsorption of additives or capping materials on the polar facets of a nanostructure, however, the way it grows can be carefully modified to control its morphology. Different unique 2D nanostructures have been produced via the introduction of different additives, such as citrate ions [271], CTAB [272] and sodium triphosphate (STP) [273], into the hydrothermal system. 2D nanostructures, exposing different facets than those the 1D NWs usually do, provide a great chance to study the impact of the different exposed facets on the functional properties of the nanostructures. Sun *et al.* [274] reported a room-temperature multistage hydrothermal synthesis of 3D flowerlike ZnO superstructures consisting of many intersecting nanosheets (Figure 4.4). The morphology of the grown structures showed high dependence on the concentrations of NaOH and trisodium citrate dihydrate. The 3D flowerlike ZnO superstructures were further optically

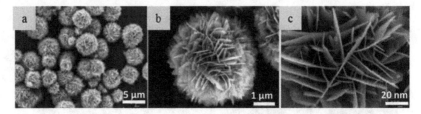

Figure 4.4 SEM images at different magnifications ((a) low and (b and c) high magnification). (Reprinted with permission from [274].)

characterised, and the optical properties were found to be significantly affected by the change in morphology.

The 3D flowerlike ZnO superstructures were tested as a photocatalyst material and demonstrated high performance, which the authors attributed to the increased surface-to-volume ratio, excellent optical properties and unique flowerlike superstructures [274], The building blocks of the 3D flowerlike superstructures reported are 2D nanosheets, which have differently exposed surface structures than that of the usually shown NW building blocks. The different exposed facets should have a significant impact on the performance of the nanostructures especially when used as a photocatalyst. Hierarchical ZnO NWs or nanoforests, to be applied in the fabrication of more-efficient dye-sensitised solar cells, were presented by Ko *et al.* [275, 276]. The growth process has two growth habits based on the growth conditions as represented schematically in Figure 4.5. The first is the axial growth habit, which increases the length of the initial NWs. The other growth habit is the lateral growth of the secondary NWs or the nanobranches. In Figure 4.5, stage (a) represents the seeding process, and stage (b) represents the growth of the initial NWs using a mixture solution of zinc nitrate and HMTA.

Stage (c) represents the possibility to extend the length of the NWs by repeating stage (b) several times. Repeating stage (b) after heating the initial NWs and adding seeds to the growth solution starts growth of secondary NWs or nanobranches, which is the stage depicted in (d). Repeating stage (d) several times increases the length of the secondary NWs and starts new nanobranches in sequence. It is clear that this process will lead to nanostructures with ultrahigh surface-to-volume ratio. Zhang *et al.* [277], reported a well-controlled synthesis technique to produce hierarchical ZnO nanostructures (Figure 4.6) via careful tuning of citrate and diaminopropane (DAP). This report investigated the effects of using these two materials on the morphological properties of the final products in multistage processes. ZnO nanostructures resulting from

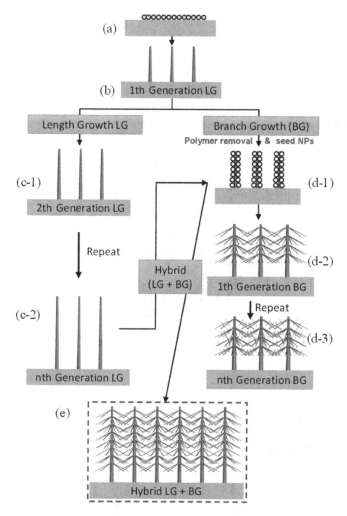

Figure 4.5 Schematic of the two possible growth processes of nanoforests: axial growth (a, b and c) and lateral growth (a, b, c and d), and (e) represents a hybrid growth process. (Reprinted with permission from [276].)

the second and third growth processes using initial NWs depended greatly on the order of adding these two materials.

As schematically presented in Figure 4.6(a), adding DAP first in the secondary growth process resulted in the growth of secondary NWs on all the six side facets of the initial NWs. Further, addition of citrate to that resulted in the tertiary growth of nanoplates from the secondary NWs. Alternatively, switching

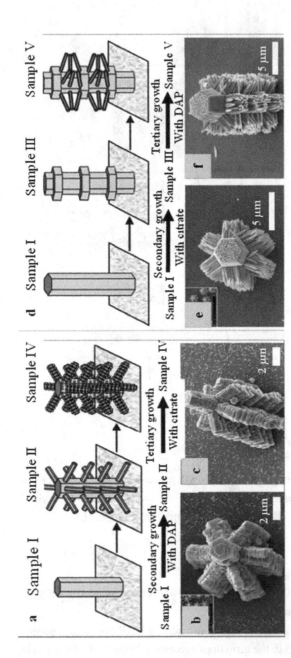

Figure 4.6 (a) Schematic of the role of adding DAP and citrate in the different growth process, and (b) top and (c) tilted view SEM images of the hierarchical ZnO nanostructure. (d) Schematic of role of adding citrate followed by diaminopropane on the growth process, and (e) top and (f) tilted view SEM images of hierarchical ZnO nanostructures. (Reprinted with permission from [277].)

the order of these two capping agents resulted in the nanostructures shown in Figure 4.6(d)–(f) [277].

4.3 Controlled Hydrothermal Synthesis

There are many factors that may greatly impact the functional properties of ZnO such as its morphology, aspect ratio, size and exposed facets. Subsequently, the performance of ZnO nanostructured devices in any application will be affected by these factors. Understanding the dependence of the functional properties on the morphology is essential to improve the performance of these devices. Therefore, the development of morphology-controlled synthesis is of great importance to enhance the performance of ZnO nanostructured devices.

4.3.1 Seed Layer–Assisted Synthesis

The ZnO NW arrays shown in Figure 4.7 can be produced placing a seeded Si substrate in a growth solution consisting of 10 mM zinc nitrate and 5 mM HMTA at 90°C for 6 hours. The NWs grown on the seed layer show high alignment and uniformity and are about 1.8 µm in length with an average diameter of 250 ± 35 nm, while the NWs grown in the bulk solution have an average length of 4.5 ± 0.6 µm and an average diameter of 540 ± 110 nm. Figure 4.7(c) shows the tip of a single ZnO NW grown in the solution where the uniform hexagonal shape is prominent. In this growth system, the solution was supersaturated with respect to zinc hydroxide, which caused most of the nutrient to react immediately and form ZnO solids homogeneously in the solution. These excess ZnO structures grown in the solution can easily contaminate the growth on the substrate, as can be clearly seen in Figure 4.7, which jeopardises device performance in many applications.

Figure 4.7 (a) Top, (b) cross-sectional and (c) edge view SEM images of ZnO NWs grown using HMTA without ammonium hydroxide [278].

Many applications require NWs with high aspect ratio, which can be achieved via extending the length of the grown NWs. It was reported, however, that it is practically challenging to hydrothermally produce ZnO NWs with a length exceeding 10 μm [243, 244]. Using a solution mixture of zinc nitrate and HMTA in water similar to the growth conditions mentioned above, the length of NWs can be extended only by refreshing the growth solution every few hours. This is due to the fast consumption of the nutrient in such growth systems. Adding ammonium hydroxide in the growth solution can significantly prevent ZnO structures from nucleating homogeneously in the solution because ammonia tends to coordinate with Zn^{2+}, thereby lowering the number of available Zn^{2+} in the solution. This will slow the growth rate and make it difficult for homogeneous nucleation to occur. In other words, heterogeneous nucleation on seeds is preferred because it requires less activation energy compared to the homogeneous process. Hence, if the right concentration of ammonium hydroxide is provided (the amount that allows heterogeneous and prevents homogeneous nucleation in the solution), ZnO NWs can grow from the seeds and the nutrient will not be consumed swiftly [243, 245, 279]. On the other hand, it was observed that including ammonium hydroxide in the growth system slows down the growth rate because it makes the solution supersaturated and limits the number of free Zn^{2+} ions [243, 245, 279]. To solve the problem of slow growth rate while achieving high aspect ratio (long NWs with relatively small diameters) without the unwanted structures grown in the bulk solution, PEI was used to extend the length of the NWs and prevent the homogeneous growth further. Previous reports presented PEI as a material that extends the length of the NWs by inhibiting their radial growth but allowing their axial growth [243, 245, 279]. This polymer is known to prefer being adsorbed on the nonpolar sides of the crystals in the solution, stopping them from growing further along these facets [243, 245, 279].

Figure 4.8 shows different SEM images of a ZnO NW array grown on a Si substrate with a seed layer using a solution mixture of zinc nitrate, HMTA, PEI and ammonium hydroxide at 90°C for 36 h. From these images, the high density, good alignment and uniformity of the NWs are evident. Moreover, they show that the substrate is clean and free from precipitates that tend to form from bulk solution. The length and diameters of the NWs can be seen clearly in the edge view SEM images in Figure 4.8(b)–(c). The NWs are about 30 μm in length with an average diameter of 115 ± 62 nm. One can observe the thin seed layer between the NWs and the substrate. Besides the high aspect ratio of NWs and clean NW array, the most important

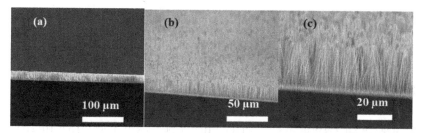

Figure 4.8 SEM images of a ZnO NW array. (a) Low-magnification cross-sectional view, (b) and (c) tilted view [278].

achievement of using this recipe at this stage is the level of control over the morphology of NWs. The diameter is almost independent of the growth time and solution concentration, as it was controlled predominantly by the PEI concentration. This suggests that extending the growth time should only increase the length and enhance the aspect ratio, while increasing the concentration of the solution should only improve the density of the NWs. In other words, every parameter will cause one effect, which gives us more rational control to examine these parameters separately and relate their impacts on the performance of the desired devices.

Moreover, growth efficiency is enhanced, as refreshing the growth solution is not required anymore because the nutrient is not consumed very fast due to the addition of ammonium hydroxide and PEI.

4.3.2 Controlling the Length of the NWs

Introducing PEI in addition to the assistance of a seed layer in the growth system liberates the density and diameter of the grown NWs from the influence of the growth time. Moreover, if the right concentration of ammonium hydroxide is used, NWs can be grown for long periods without the need of refreshing the growth solution as discussed earlier. Taking advantage of this, the longest hydrothermally grown NWs ever were produced. Also, it is possible to control the length of the NWs and accordingly the aspect ratio of the ZnO NW arrays by simply adjusting the growth time. Figure 4.9 shows cross-sectional view SEM images of arrays of ZnO NWs produced under identical growth conditions for different periods.

4.3.3 Controlling the Density of NWs

The nutrient concentration of $(Zn(NO_3)_2$ and HMTA in the growth solution maintaining the ratio of 2:1 has a great impact on the density of the grown ZnO

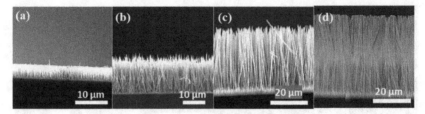

Figure 4.9 ZnO NW arrays grown for different periods of time: (a) 5 h, (b) 20 h, (c) 40 h and (d) 60 h [278].

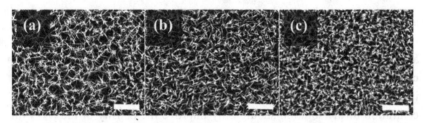

Figure 4.10 Top view SEM images of ZnO NW arrays grown with (a) 10, (b) 20 and (c) 30 mM $Zn(NO_3)_2$. (Scale bars are 2 μm [278].)

NWs. It was found that the density of NWs can be tuned by adjusting the concentration of the nutrient. Figure 4.10(a)–(c) show top view SEM images of different arrays of ZnO NWs that are grown using solutions with different nutrient concentrations (10, 20 and 30 mM, respectively). These nutrient concentrations resulted in average densities of 40.3 ± 5.5, 62.7 ± 9 and 78.3 ± 6.8 NW.μm^{-2}, respectively. As discussed earlier, since PEI is included in the growth solution, the diameter of the grown NWs will be controlled predominantly by the concentration of PEI. Therefore, it is concluded that the density of the ZnO NWs is controlled by the concentration of nutrient in the growth solution.

4.4 Seedless Hydrothermal Growth

In many reports on the hydrothermal synthesis of ZnO nanostructures, the seed layer is an essential part of the growth system [243–245]. Nevertheless, for many applications the seed layer is not preferred [279] as it usually requires annealing at a relatively high temperature prior to the growth, which is not suitable with low-cost flexible substrates [243]. Therefore, a controlled seedless and site-selective hydrothermal method for producing ZnO NWs is of interest. Additionally, attempts to control the density and morphology of the grown ZnO

NWs by adjusting the concentration of ammonium hydroxide and the nutrient concentration in the refreshing solution will also be discussed.

4.4.1 Influence of Ammonium Hydroxide Concentration

Figure 4.11 shows SEM images of ZnO NWs grown without seeds on Au-coated Si/SiO$_2$ substrates using hydrothermal synthesis with different concentrations of NH$_3$.H$_2$O at 90°C for 3 h. Without adding ammonium hydroxide, the resulted NWs had an average diameter of 200 ± 43 nm and length of about 1 ± 0.15 μm (Figure 4.11(a)). Most of the nutrient material has been consumed by nucleating homogeneously and was found at the base of the vial after the growth. The case became different once ammonium hydroxide was included in the growth solution. Figure 4.11(b)–(i) show SEM images of ZnO NWs grown without seeds at ammonium hydroxide concentrations of 0.05, 0.1, 0.15, 0.2, 0.25, 0.3, 0.35 and 0.45 M, respectively. At a concentration level of 0.05 M of ammonium hydroxide, the nucleation rate on the substrate improved significantly in comparison with the nucleation rate without ammonium hydroxide, as evident in Figure 4.11(b). The density of the NWs at this level was 12 ± 1.1 NW.μm^{-2} and the average diameter was 120 ± 45 nm. In this case the amount of NH$_4^+$ was not sufficient to consume all of the zinc hydroxide [243, 245, 279]. Consequently, at this low concentration level of ammonium hydroxide, the length of the NWs was relatively small at 300 nm (Figure 4.11(b)). Raising the concentration of ammonium hydroxide further to 0.1 M decreased the rate of releasing Zn^{2+} further and limited the available nutrient materials at the beginning of the growth process. As a result, homogeneous nucleation in the bulk solution was suppressed a bit more, encouraging the nucleation on the substrate and increasing the density, diameter and length of the NWs to 14 ± 1.25 NW.μm^{-2}, 175 ± 32 nm and 400 nm, respectively (Figure 4.11(c)). At an ammonium hydroxide concentration of 0.15 M, the density of the NWs increased slightly to 17 ± 0.95 NW.μm^{-2}. The diameter and length also were increased, to 200 ± 39 nm and 600 nm respectively (Figure 4.11(d)). Further increase in the concentration of ammonium hydroxide beyond this level caused the average diameter of the NWs to decrease because the number of free Zn^{2+} at the beginning of the growth process was getting smaller while the density of nucleation sites was increasing.

At the concentration level of 0.2 M, the average diameter of the NWs was 180 ± 37 nm, and their length was 900 nm (Figure 4.11(e)). The density was increased further with this increase in the concentration of ammonium hydroxide to 23 ± 1.3 NW.μm^{-2} as a result of the strongly encouraged nucleation on the substrate. As the concentration of ammonium hydroxide was further

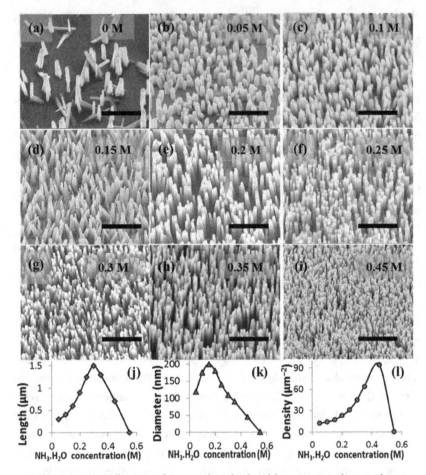

Figure 4.11 Influence of ammonium hydroxide concentration on the morphology of ZnO NWs grown on seedless gold layer. (a)–(i) SEM images of NWs grown under ammonium hydroxide concentrations of 0, 0.05, 0.1, 0.15, 0.2, 0.25, 0.3, 0.35 and 0.45 M, respectively. (The scale bar is 2 μm.) (j)–(l) Plots show dependence of length, diameter and density of the NWs on concentration of ammonium hydroxide [278].

increased to 0.25, 0.3, 0.35 and 0.45 M, the density of the NWs kept increasing to 31 ± 1, 45 ± 0.9, 64 ± 0.7 and 94 ± 1.2 NW.μm^{-2}, respectively. The diameter of the NWs on the other hand kept decreasing to 140 ± 33, 110 ± 49, 90 ± 52 and 48 ± 31, respectively. The following length values of 1.25, 1.5, 1.3 and 0.7 μm were measured respectively for the same levels of ammonium hydroxide concentration. To clearly observe and analyze the impact of ammonium hydroxide concentration on the morphology of the grown NWs, different areas of the substrates were analysed, and the length, average diameter and average density of the as-grown NWs were calculated and represented in the plots. The plots in Figure 4.11(j)–(l) represent the length, average diameter and average density of the as-grown NWs as functions of the concentration of ammonium hydroxide, respectively. The length of the NWs increases with the increase in concentration of ammonium hydroxide from 0.05 to 0.25 M where the maximum value occurs and then starts to decrease, as shown in Figure 4.11 (j). For the average diameter of the NWs, the same trend is observed with the maximum value being at 0.15 M. For the ammonium hydroxide concentration range from 0.05 to 0.45 M, the density of the NWs kept increasing and the maximum was attained at 0.45 M. The effect of the increasing density of NWs on their diameter size and length is understood since the amount of nutrient is fixed in the solution and, as a result, increasing the nucleation sites leads to smaller NWs. By further increasing the concentration of ammonium hydroxide up to 0.55 M, the huge amount of NH_4^+ prevented any nucleation on the substrate, and no NWs were grown.

4.4.2 Selective Seedless Growth on Au Surfaces

It is shown that the concentration of ammonium hydroxide controls the length, diameter and density of the NWs. The nucleation process can also be controlled by optimising the concentration of ammonium hydroxide to reach the point where the growth position can be chosen. In other words, the growth process becomes site selective (Figure 4.12). Due to the smaller lattice mismatch between Au and ZnO [265], ZnO NWs were grown on substrates with different prepatterns of Au, which are shown in Figure 4.12. It is evident in these SEM images that ZnO NWs grew significantly on areas deposited with Au, while no

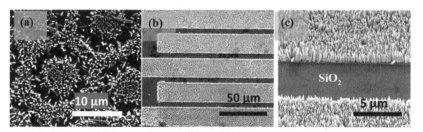

Figure 4.12 (a), (b) and (c) SEM images of the prepared ZnO NWs on Si/SiO$_2$ substrates with different prepatterned Au electrodes at different magnifications [278].

nucleation on silicon dioxide can be seen. These results clearly show that the aligned ZnO NWs can grow selectively on Au. This synthesis method can be used in many applications, such as gas sensors and photodetectors.

4.4.3 Influence of the Refreshing Growth Solution

The density of ZnO NWs can be improved greatly by increasing the concentration of ammonium hydroxide with the observed low growth rate. With the intention of enhancing the aspect ratio of the grown NWs, PEI was used to extend the length of the NWs by inhibiting the radial growth but allowing the axial growth. Despite all the great results achieved by using PEI in the seed layer–assisted growth method, it was observed that introducing PEI in the seedless synthesis system at the beginning of the growth process and before the initial nucleation stage disturbs the entire growth process. In the nucleation stage, PEI will encapsulate any formed nuclei. This observation has been reported previously [279]. Hence, in order to overcome this problem, PEI was introduced after the nucleation stage completed by refreshing the growth solution with a new one containing PEI.

(a) Nutrient Concentration in the Refreshing Growth Solution

After refreshing the growth solution, the density of NWs can be controlled by simply tuning the nutrient concentration in the refreshing solution. The diameter and length of the grown NWs will be controlled predominantly by the amount of PEI added in the refreshing solution and the growth time, respectively. Figure 4.13 shows tilted view SEM images of ZnO NWs grown without seed on Au-coated substrates using refreshing growth solution of different nutrient concentrations. The initial growth solution was a mixture of zinc nitrate (25 mM), HMTA (12.5 mM) and ammonium hydroxide (0.35 M).

In the refreshing solution, the concentrations of PEI (5 mM) and ammonium hydroxide of (0.35 M) were fixed, and the concentrations of $Zn(NO_3)_2$ and HMTA (maintaining the same ratio of 2:1) were varied. Therefore, for this set of experiments, the concentration of zinc nitrate alone represents the nutrient concentration. In the first experiment, when only PEI and ammonium hydroxide were added as a refreshing solution without zinc nitrate or HMTA, the growth process stopped completely, and no growth occurred after adding the refreshing solution due to the absence of the nutrient materials (Figure 4.13(a)). A low-density growth, however, was observed when a 3 mM solution of zinc nitrate was added, as shown in Figure 4.13(b). The concentration of the nutrient in the refreshing growth solution was further increased to 6, 10, 15 and 25 mM, improving the density accordingly, as evident in Figure 4.13(c)–(f), respectively.

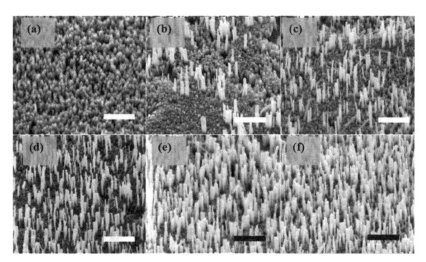

Figure 4.13 Tilted view SEM images of ZnO NWs grown without seed on Au-coated substrates using refreshing growth solutions with a $Zn(NO_3)_2$ concentration of (a) 0 mM, (b) 3 mM, (c) 6 mM, (d) 10 mM, (e) 15 mM, and (f) 25 mM. (Scale bars are 2 μm [278].)

(b) PEI Concentration in the Refreshing Growth Solution

The influence of PEI concentration in the refreshing solution on the grown ZnO NWs was investigated by fixing all other parameters. Initially, less than 3 mM of PEI was added to the refreshing solution, and after 2 h the NW length was extended from 200 to 400 nm with no significant change observed in the diameters or density of the NWs. When 5 mM of PEI was added, the length of the NWs increased from 200 to 800 nm with a limited effect on their diameters and density (Figure 4.14(a)). Increasing the concentration of added PEI up to 10 mM, however, increased the length of the NWs from 0.2 to 1.1 μm and caused a dramatic reduction in the diameters of the NWs from 150 to ~25 nm or less (Figure 4.14(b)). This significant change in the diameter of the NWs after refreshing the growth solution transformed the NWs into syringe-like nanostructures or nanosyringes (NSs), as shown in Figure 4.14(c)–(f). Increasing the PEI concentration up to 15 mM led to a total encapsulation of the NWs, and no further growth occurred.

4.4.4 Hierarchical Nanowires

Hierarchical nanostructures assembled from 1D and 2D nanostructure building blocks possess exceptional physical and chemical properties and have recently attracted interest [280, 281]. Such structures are investigated for many possible applications, such as sensors [282], photocatalysis [283–285], fuel cells [286] and

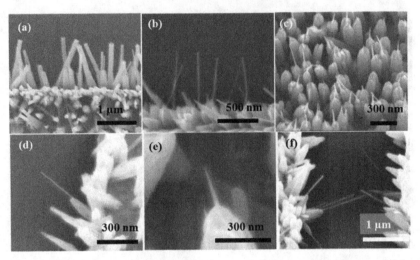

Figure 4.14 SEM images of (a), (b) ZnO nanosyringes grown at the edge of Au electrode, (c) array of ZnO nanosyringes, (d, e) the needle of the nanosyringe and (f) two ZnO nanosyringes bridging a gap between two Au electrodes [278].

drug release systems [287]. Regardless of all the advantages of the solution-phase synthesis methods in producing monomorphological structures like nanoparticles, NWs, nanorods and NDs, it is still a challenge to use these strategies in the production of hierarchical architectures. In this section, a well-controlled multistage hydrothermal technique to produce hierarchical ZnO NWs (HZNWs) is introduced. The hierarchical nanostructures produced in this simple growth method are rationally designed by adjusting the growth process parameters. SEM images obtained at different magnifications of HZNWs assembled from initial 1D ZnO NWs (seeds in this case) are shown in Figure 4.15. These SEM images show clearly that the secondary NWs organise themselves into orderly arrays forming a 6-fold symmetry. By looking closely at the morphology of the HZNWs it can be observed that the secondary NWs grow on the side facets of the initial ZnO NW. The hexagonal shape of the initial ZnO NWs caused the 6-fold symmetry of the secondary NWs [289]. This form of structure was reported widely in the literature [281, 289–292]. As mentioned previously, in hydrothermal synthesis, seed-assisted nucleation is much easier than nucleating in the solution without seeds. Hence, initial 1D ZnO nanostructures can serve as seeds to have the secondary NWs grow on them. In a search for more control over the morphology of the grown hierarchical ZnO structures, investigations have been conducted to identify the role of each reaction parameter involved in the growth process. The investigations carried out show that the key parameters in the growth process are

Figure 4.15 SEM image of (a) many HZNWs at low magnification, (b) single HZNW with 6-fold symmetry, (c) single HZNW at high magnification, (d) side view of a single HZNW and (e) and (f) top view of HZNW with 6-fold symmetry [278].

the morphology of the initial ZnO nanostructures/seeds, the nutrient concentration in the growth solution, growth time and the amount of the PEI added in the secondary stage of the growth. Therefore, by understanding the role of each of these parameters, one can achieve the optimum conditions to produce hierarchical ZnO nanostructures that can meet the requirements of a specific application.

(a) Influence of Initial ZnO Seeds/Nanostructures

Understanding the influence of the ZnO nanostructures/seeds morphology is very important for controlling the growth and properties of the produced structures. These seeds will be the key platform which gives the overall shape to final structure. In our experiments, hierarchical ZnO nanostructures can be synthesised using nanostructures/seeds with a variety of different morphologies, as seen in Figure 4.16(a)–(c) [278]. Moreover, the morphology of the secondary ZnO NWs is not affected by the polarity of the exposed facets of initial nanostructure on which they are grown (i.e., secondary ZnO NWs grow equally on polar and nonpolar facets). The density and uniformity of the grown secondary ZnO NWs, however, depend significantly on the space available on the initial ZnO nanostructure. Figure 4.16(d)–(f) show SEM images of

Figure 4.16 (a), (b) and (c) Hierarchical ZnO structures grown using initial ZnO nanostructures with different morphologies; HZNWs grown using ZnO NWs with average diameter of (d) 100, (e) 300 and (f) 500 nm [278].

HZNWs grown at the same growth conditions using initial ZnO NWs with different diameters (100, 300 and 500 nm, respectively). For the initial ZnO NW with the smallest diameter (100 nm), there is only one row of secondary NWs on each side surface of the core ZnO NW, as evident in Figure 4.16(d). This could be observed in the encircled regions of the SEM image (Figure. 4.16d).

(b) Influence of Nutrient Concentration on Growth

The nutrient concentration in growth solution has a significant influence on the density of the grown secondary NWs and has a limited effect on their diameter. The nutrient concentration has a similar impact in both cases when 1D and 2D ZnO nanostructures are used as seeds. The effect of changing the growth solution nutrient concentration is more noticeable in the HZNWs. The reason is that the 6-fold symmetry structure of the HZNWs, which is easy to notice, is very sensitive to the nutrient concentration, and it can be achieved only under optimised concentrations of the growth solution. Figure 4.17 shows SEM images of HZNWs grown at different concentrations of the growth solution.

When the growth-solution concentration was 5 mM, the density of the secondary NWs was low (Figure 4.17(a)). When the concentration of the growth solution increased between 10 and 20 mM, the 6-fold symmetry in the solution concentration was further increased to values above 25 mM and the number of the secondary

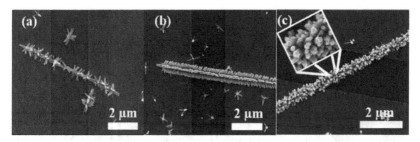

Figure 4.17 (a) HZNWs grown with growth-solution concentration of 5 mM, (b) 12.5 mM and (c) 25 mM [278].

NWs also increased, forming HZNWs with high-density branches (Figure 4.17(c)). It is obvious in the SEM image (Figure 4.17(c)) that the HZNWs have lost their 6-fold symmetry. In this hierarchical nanostructure, the surface-to-volume ratio has increased as well as the fraction of exposed polar facets in comparison with the HZNWs grown at lower growth-solution concentrations. Hierarchical nanostructures can be clearly seen (Figure 4.17(b)). The nutrient concentration in the growth solution is essential and controls the nucleation rate of the secondary NWs. At low concentrations, only a few disordered secondary NWs grew on the most favorable nucleation sites on the surface of the initial ZnO NWs. Increasing the nutrient concentration in the growth solution raised the nucleation rate and increased the density of the secondary NWs while maintaining the alignment and the 6-fold symmetry. Increasing the concentration beyond this value pushed the nucleation rate further, causing the secondary NWs to grow almost everywhere. Subsequently, the HZNWs lost their 6-fold symmetry but increased the surface-to-volume ratio and fraction of exposed polar facets.

(c) Influence of Growth Time on Structural Morphology

The effect of the growth time was investigated by growing HZNWs for different periods of time. Figure 4.18(a) and (b) show SEM images of HZNWs grown using a 12.5 mM solution for 2 and 5 h, respectively. From the SEM images it can be seen that the length of the secondary ZnO NWs was extended from 0.5 to 2 μm, with no obvious change in their diameter size or density. Due to the presence of PEI in the growth reaction system, the diameter is almost independent of all other factors in the reaction, as discussed in the previous section. To make sure that these observations made on the influence of growth time are independent of the concentration of the growth solution, the experiments were repeated with different concentrations. SEM images of HZNWs grown with 25 mM growth solution concentration for 2 and 5 h are shown in Figure 4.18(c)–(d), respectively. It is

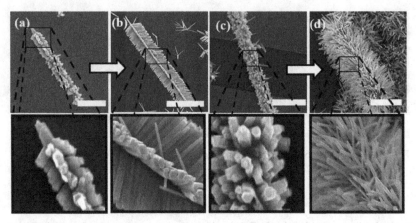

Figure 4.18 HZNWs grown with (a) 12.5 mM for 2 h, (b) 12.5 mM for 5 h, (c) 25 mM for 2 h and (d) 25 mM for 5 h. (Scale bars are (a) 2, (b) 5, (c) 2 and (d) 5 μm [278].)

evident from the above SEM analysis that adjusting the growth time is an effective tool to control the length of the grown secondary NWs in the HZNW structures without affecting their density or diameter. This is understood from the fact that the secondary-growth NWs are covered with PEI from all exposed nonpolar facets except the top polar (0001) facet.

5 Suitability of Inorganic NWs for Large-Area Flexible Electronics

The NWs offer attractive opportunities for large-area flexible electronics, as unlike bulk they could be printed on various substrates to obtain ultrathin electronic layers. Unlike thin films, which can be grown at low temperatures directly on the flexible substrates, the NW-based electronic layers offer better quality, owing to the high crystallinity of these nanostructures. For example, the mobility of thin-film amorphous Si is ~1 cm^2 V^{-1} s^{-1} and that of electronic layer with printed array of InAs NWs is 92 cm^2 V^{-1} s^{-1} [57]. Further, the uniformity of electronic layers is likely to be better with printing than direct deposition of thin films. As mentioned in the introduction section, however, the NW approach for printing electronic layers requires alternative methods such as contact and transfer printing, which are briefly discussed below.

5.1 Contact Printing

Contact printing has been established as a direct transfer process of NWs from high-temperature substrates such as Si, sapphire, metal, etc. to flexible substrates at room temperature [32, 43, 53]. Contact printing eliminates all intermediate steps (such as solution dispersion), which affects the quality of synthesised NWs, and it is advantageous for all bottom-up grown semiconducting NWs such as Si, GaAs, GaN, etc. The typical NW print set-up consists of a precisely controlled vertical stage where the high-temperature NW substrate (donor) is firmly fixed (Figure 5.1(b))[30]. The flexible receiver substrate is attached over a horizontal linear stage, which makes a guided contact with donor substrate for NW transfer. The transfer printing process is centered on directional sliding of NW substrate (donor) over a receiver flexible polymer substrate, which eventually leads to the formation of a horizontal array of NWs. The controlled shear force at the donor–receiver interface helps to align the NWs along the sliding direction. The NW transfer yield and alignment is improved by surface functionalization using organic lubricants. These agents anchor the detached NWs from the donor substrates and aids aligned printing by reducing NW–NW friction in the printing process. The selective surface functionalization is a room-temperature process, which is compatible with flexible substrates such as polyamide, PVC, etc. Recently, planar NW heterostructures have been demonstrated using a contact printing process where alternating layers of Si and ZnO NWs are printed over flexible PVC substrates (Figure 5.1(d)–(e))[32]. The transparent flexible NW circuit has been demonstrated for UV photodetector application (Figure 5.1(e)). This shows that ZnO and Si NWs grown at high temperatures and subsequently fabricated over flexible substrates can be used for applications in flexible electronics. The contact printing process has the potential to scale up for a NW batch transfer process. Differential roll printing is one variant of contact printing to scale up for large-area, roll-to-roll (R2R) printing (Figure 5.1(f)–(g)) [75]. NWs grown on solid cylinders are used as donor in the printing process. Many research groups are currently investigating R2R printing to scale up the printing process. In essence, contact printing seamlessly integrates the high-temperature NW growth process with low-temperature device fabrication over flexible substrates and is evolving with batch production possibility.

5.2 Transfer Printing

Transfer printing or stamping techniques are based on physical transfer of horizontal 1D nanostructures fabricated by top-down techniques over semiconducting wafers such as Si, GaAs, Ge, etc. [23, 26, 48, 56].These dry transfer processes seamlessly transfer 1D array or fabricated devices from inorganic wafers to flexible plastic substrates. The top-down fabrication process is carried out either by lithography (e-beam or optical) followed by dry plasma etching or by wet chemical etching using strong acids or bases [292]. Top-down

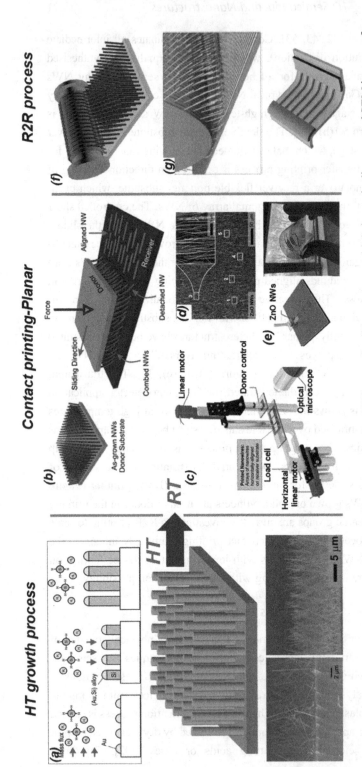

Figure 5.1 Process flow of the contact printing of bottom-up grown NWs over flexible substrates. (a) Schematic diagram and SEM images of Si and ZnO NWs grown by VLS method. (b) Schematic illustration of planar contact printing process (c) Schematic diagram of indigenously built contact printing set-up. (d) SEM image of contact-printed, aligned, highly dense ZnO NWs. (e) Illustration and photograph of flexible UV photodetector using contact printing of Si and ZnO NWs. (f, g) Schematic illustration of roll-to-roll contact printing process. (Reprinted with permission from [32].)

techniques help to realise high-aspect-ratio 1D nanostructures (ribbons, wires, belts) with the smallest dimension in the order of <10 nm and length in the order of >100 μm (Figure 5.2(a)–(b)). Parent wafers such as SOI and Gaas/AlGaAs were typically used to produce nanoribbons and FETs of Si and Gaas/AlGaAs 1D heterostructures. In addition, the device fabrication involves dopant diffusion at high temperature and oxidation (dry, wet and spin on doping). These top-down etching and fabrication processes are not compatible with flexible substrates. The etched 1D structures are weakly attached with anchoring points to the source wafer. A flexible PDMS stamp was made to contact with the 1D structures array and gently peeled off from the surface [175]. The 1D array subsequently transferred over an epoxy-coated plastic substrate (Figure 5.2(c)). The transfer process has led to highly dense large-area flexible Si FETs, which has applications in gas sensing (Figure 5.2(d)–(f)). The transfer printing process bridges high-performance device fabrication and flexible substrates using a controllable transfer strategy.

6 Summary and Conclusions

This Element elaborately explained the growth mechanism and synthesis techniques for three technologically important types of NWs used in flexible electronics: (1) Si, (2) CNTs and (3) ZnO. The discussion is, in fact, representative of nanostructures from a wide variety of semiconducting materials. Si NW growth is explained using the well-established VLS growth mechanism and binary phase diagrams, which play a crucial role in deciding the growth of NWs. Further, the growth kinetics of the NWs has been explained using an atomistic model, which could lead to new studies involving prediction of growth rates. Likewise, the CNT growth and characterization methods have been explained using well-established techniques such as arc discharge method, laser ablation and CVD. The experimental conditions to obtain single-wall and multiwall nanotubes have been described as well as strategies to control the metallic and semiconducting CNTs. The growth mechanism and experimental techniques for ZnO NWs show how it is possible to rationally control the length and density of the grown ZnO NWs by adjusting the growth time and nutrient concentration in the growth solution, respectively. In fact, similar studies are also presented for CNTs and Si NWs. In essence, the Element has discussed intricate details in growth mechanisms, and the fundamental discussion presented here is expected to support further understanding of the relationship between growth conditions and resulting functional properties of these nanostructures. The discussion could be very well extended towards other important semiconductors such as GaAs, GaN, InAs, InGaAs, etc. as well as heterostructures.

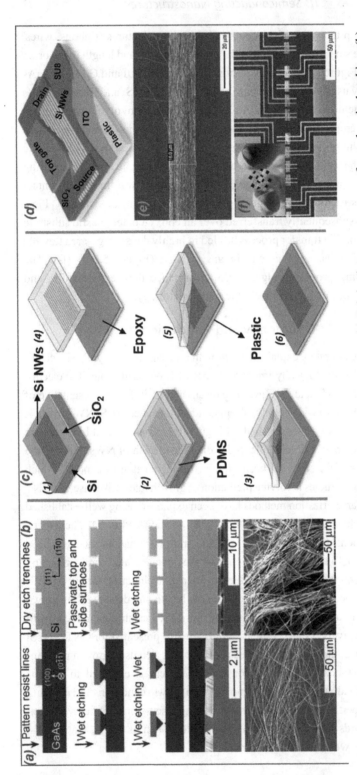

Figure 5.2 Transfer printing process of top-down fabricated 1D nanostructures. (a) Fabrication process steps of wet chemical etching. (b) Top-down dry etching process sequence and SEM images. (c) Schematic illustration of dry stamping transfer process. (d) Schematic diagram of the flexible Si NW FETs device structure. (e) SEM images of the Si NWs transferred by transfer printing technique. (f) Si NW array–based flexible gas sensor (inset photograph). (Reprinted with permission from [175, 292].)

From the point of view of flexible and large-area electronics, it is clear that due to thermal budget issues the current synthesis techniques presented in this Element do not allow direct synthesis of nanostructures on nonconventional flexible substrates (e.g., plastic). With some tools, such as photothermal CVD (PTCVD) where substrate temperature during synthesis can be between 300 and 450°C, it is possible to grow NWs directly on limited flexible substrates (e.g., polyimide and insulated metal foils). This range of temperatures, however, is still higher for several plastic substrates, and therefore indirect methods such as contact and transfer printing have been explored. These methods decouple the nanostructure growth process (which requires high temperatures) from the electronic device fabrication on flexible substrates. This means the synthesis of the NWs could still be achieved with conventional vapour-phase methods on substrates which can withstand high temperatures (e.g., Si, SiC, sapphire). Further, with addition of steps after synthesis, it is possible to transfer the NWs onto flexible substrates, and the rest of the low-temperature processing steps (e.g., metallization by printing or lithography) can be carried out to obtain electronic devices. This is an interesting direction, as it allows the quality of nanostructures to be controlled with the various tools described in this Element.

Abbreviations

0D	Zero-dimensional
1D	One-dimensional
2D	Two-dimensional
NW	Nanowire
VLS	Vapour–liquid–solid
VSS	Vapour–solid–solid
V–L	Vapour–liquid
V–S	Vapour–solid
CVD	Chemical vapour deposition
PVD	Physical vapour deposition
MOCVD	Metal organic chemical vapour deposition
LPCVD	Low pressure chemical vapour deposition
APCVD	Atmospheric pressure chemical vapour deposition
UHV	Ultrahigh vacuum
TEM	Transmission electron microscope
SEM	Scanning electron microscope
LL	Layer by layer
ML	Multilayer
RI	Rough interface
TPB	Triple phase boundary
OAG	Oxygen assisted growth
H	Gas flow rate
IF	Interface
CNTs	Carbon nanotubes
MWCNTs	Multiwall carbon nanotubes
SWCNTs	Single-wall carbon nanotubes

References

[1] G. Hodes, "When small is different: some recent advances in concepts and applications of nanoscale phenomena," *Adv. Mater.*, vol. 19, no. 5, pp. 639–655, 2007.

[2] V. T. Liveri, *Controlled synthesis of nanoparticles in microheterogeneous systems*. Springer Science & Business Media, 2006.

[3] C. Guozhong, *Nanostructures and nanomaterials: synthesis, properties and applications*. World Scientific, 2004.

[4] Y. Xia, *et al.*, "One-dimensional nanostructures: synthesis, characterization, and applications," *Adv. Mater.*, vol. 15, no. 5, pp. 353–389, 2003.

[5] C. M. Lieber, "One-dimensional nanostructures: chemistry, physics & applications," *Solid State Commun.*, vol. 107, no. 11, pp. 607–616, 1998.

[6] P. Yang, "The chemistry and physics of semiconductor nanowires," *MRS Bull.*, vol. 30, no. 2, pp. 85–91, 2005.

[7] O. Hayden, R. Agarwal, and W. Lu, "Semiconductor nanowire devices," *Nano Today*, vol. 3, no. 5–6, pp. 12–22, 2008.

[8] Y. Li, F. Qian, J. Xiang, and C. M. Lieber, "Nanowire electronic and optoelectronic devices," *Mater. Today.*, vol. 9, no. 10, pp. 18–27, 2006.

[9] F. Patolsky, G. Zheng, and C. M. Lieber, "Nanowire-based biosensors," ed: ACS Publications, Anal. Chem, Vol. 78, no. 13, pp. 4260–4269, 2006.

[10] X. Chen, C. K. Wong, C. A. Yuan, and G. Zhang, "Nanowire-based gas sensors," *Sens. Actuator B-Chem.*, vol. 177, pp. 178–195, 2013.

[11] Y. Qi and M. C. McAlpine, "Nanotechnology-enabled flexible and biocompatible energy harvesting," *Energ. Environ. Sci.*, vol. 3, no. 9, pp. 1275–1285, 2010.

[12] F.-R. Fan, Z.-Q. Tian, and Z. L. Wang, "Flexible triboelectric generator," *Nano Energy*, vol. 1, no. 2, pp. 328–334, 2012.

[13] E. C. Garnett, M. L. Brongersma, Y. Cui, and M. D. McGehee, "Nanowire solar cells," *Ann. Rev. Mater. Res.*, vol. 41, pp. 269–295, 2011.

[14] C. K. Chan, X. F. Zhang, and Y. Cui, "High capacity Li ion battery anodes using Ge nanowires," *Nano Lett.*, vol. 8, no. 1, pp. 307–309, 2008.

[15] R. Saito, G. Dresselhaus, and M. S. Dresselhaus, *Physical properties of carbon nanotubes*. World Scientific, 1998.

[16] M. M. Shulaker, *et al.*, "Carbon nanotube computer," *Nature*, vol. 501, no. 7468, p. 526, 2013.

[17] C. Rao and A. Govindaraj, "Synthesis of inorganic nanotubes," *Adv. Mater.*, vol. 21, no. 42, pp. 4208–4233, 2009.

[18] J. Goldberger, R. Fan, and P. Yang, "Inorganic nanotubes: a novel platform for nanofluidics," *Accounts Chem. Res.*, vol. 39, no. 4, pp. 239–248, 2006.

[19] L. J. Lauhon, M. S. Gudiksen, D. Wang, and C. M. Lieber, "Epitaxial core–shell and core–multishell nanowire heterostructures," *Nature*, vol. 420, no. 6911, pp. 57–61, 2002.

[20] K. Takei, *et al.*, "Nanowire active-matrix circuitry for low-voltage macroscale artificial skin," *Nat. Mater.*, vol. 9, no. 10, pp. 821–826, 2010.

[21] M. Paladugu, *et al.*, "Novel growth phenomena observed in axial InAs/GaAs nanowire heterostructures," *Small*, vol. 3, no. 11, pp. 1873–1877, 2007.

[22] N. P. Dasgupta, *et al.*, "25th anniversary article: semiconductor nanowires–synthesis, characterization, and applications," *Adv. Mater.*, vol. 26, no. 14, pp. 2137–2184, 2014.

[23] Y. Sun and J. A. Rogers, "Inorganic semiconductors for flexible electronics," *Adv. Mater.*, vol. 19, no. 15, pp. 1897–1916, 2007.

[24] C. M. Lieber and Z. L. Wang, "Functional nanowires," *MRS Bull.*, vol. 32, no. 2, pp. 99–108, 2007.

[25] Z. Liu, J. Xu, D. Chen, and G. Shen, "Flexible electronics based on inorganic nanowires," *Chem. Soc. Rev.*, vol. 44, no. 1, pp. 161–192, 2015.

[26] Y. Sun and J. A. Rogers, *Semiconductor nanomaterials for flexible technologies: From photovoltaics and electronics to sensors and energy storage*. William Andrew, 2010.

[27] M. R. Alenezi, S. J. Henley, N. G. Emerson, and S. R. P. Silva, "From 1D and 2D ZnO nanostructures to 3D hierarchical structures with enhanced gas sensing properties," *Nanoscale*, vol. 6, no. 1, pp. 235–247, 2014.

[28] B. O. Boskovic, V. Stolojan, R. U. Khan, S. Haq, and S. R. P. Silva, "Large-area synthesis of carbon nanofibres at room temperature," *Nat. Mater.*, vol. 1, no. 3, pp. 165–168, 2002.

[29] C. Garcia Nunez, W. T. Navaraj, F. Liu, D. Shakthivel, and R. Dahiya, "Large-area self-assembly of silica microspheres/nanospheres by temperature-assisted dip-coating," *ACS Appl. Mater. Inter*, vol. 10, no. 3, pp. 3058–3068, 2018.

[30] F. L. C. G. Nunez, W. T. Navaraj, A. Christou,D. Shakthivel, and R. Dahiya, "Heterogeneous integration of contact-printed semiconductor nanowires for high performance devices on large areas," *Microsyst Nanoeng*, 2018.

[31] F. L. C. G. Núñez, S. Xu and R. Dahiya, "Large-area electronics based on micro/nanostructures and the manufacturing technologies," *Cambridge Elements (2018)*, In press.

[32] C. G. Núñez, F. Liu, W. T. Navaraj, A. Christou, D. Shakthivel, and R. Dahiya, "Heterogeneous integration of contact-printed semiconductor nanowires for high-performance devices on large areas," *Microsyst Nanoeng*, vol. 4, no. 1, p. 22, 2018.

[33] C. G. Nunez, W. Taube, F. Liu, and R. Dahiya, "ZnO nanowires based flexible UV photodetectors for wearable dosimetry," in *SENSORS, 2017 IEEE*, 2017, pp. 1–3: IEEE.

[34] D. Striakhilev, A. Nathan, Y. Vygranenko, P. Servati, C.-H. Lee, and A. Sazonov, "Amorphous silicon display backplanes on plastic substrates," *J. Disp. Technol.*, vol. 2, no. 4, pp. 364–371, 2006.

[35] A. Nathan, *et al.*, "Flexible electronics: the next ubiquitous platform," *Proc. IEEE.*, vol. 100, no. Special Centennial Issue, pp. 1486–1517, 2012.

[36] L. Petti, *et al.*, "Metal oxide semiconductor thin-film transistors for flexible electronics," *Appl. Phys. Rev.*, vol. 3, no. 2, p. 021303, 2016.

[37] R. Sporea, M. Trainor, N. Young, J. Shannon, and S. Silva, "Source-gated transistors for order-of-magnitude performance improvements in thin-film digital circuits," *Sci. Rep.*, vol. 4, 2014.

[38] S. J. Kim, K. Choi, B. Lee, Y. Kim, and B. H. Hong, "Materials for flexible, stretchable electronics: graphene and 2D materials," *Ann. Rev. Mater. Res.*, vol. 45, pp. 63–84, 2015.

[39] D. Akinwande, N. Petrone, and J. Hone, "Two-dimensional flexible nanoelectronics," *Nat. Commun.*, vol. 5, p. 5678, 2014.

[40] C. García Núñez, W. T. Navaraj, E. O. Polat, and R. Dahiya, "Energy autonomous flexible and transparent tactile skin," *Adv. Funct. Mater*, vol. 27, no. 18, 2017.

[41] E. O. Polat, O. Balci, N. Kakenov, H. B. Uzlu, C. Kocabas, and R. Dahiya, "Synthesis of large area graphene for high performance in flexible optoelectronic devices," *Sci. Rep.*, vol. 5, p. 16744, 2015.

[42] C. Wang, *et al.*, "Extremely bendable, high-performance integrated circuits using semiconducting carbon nanotube networks for digital, analog, and radio-frequency applications," *Nano Lett.*, vol. 12, no. 3, pp. 1527–1533, 2012.

[43] P. H. Lau, *et al.*, "Fully printed, high performance carbon nanotube thin-film transistors on flexible substrates," *Nano Lett.*, vol. 13, no. 8, pp. 3864–3869, 2013.

[44] S. Khan, R. S. Dahiya, and L. Lorenzelli, "Flexible thermoelectric generator based on transfer printed Si microwires," in *Solid State Device Research Conference (ESSDERC), 2014 44th European*, 2014, pp. 86–89: IEEE.

[45] S. Khan, N. Yogeswaran, W. Taube, L. Lorenzelli, and R. Dahiya, "Flexible FETs using ultrathin Si microwires embedded in solution processed dielectric and metal layers," *J. Micromech. Microeng.*, vol. 25, no. 12, p. 125019, 2015.

[46] S. Khan, L. Lorenzelli, and R. Dahiya, "Flexible MISFET devices from transfer printed Si microwires and spray coating," *IEEE. J. Electron. Devi.*, vol. 4, no. 4, pp. 189–196, 2016.

[47] W. Dang, V. Vinciguerra, L. Lorenzelli, and R. Dahiya, "Printable stretchable interconnects," *Flex. Print. Electron.*, vol. 2, no. 1, p. 013003, 2017.

[48] R. S. Dahiya, A. Adami, C. Collini, and L. Lorenzelli, "Fabrication of single crystal silicon micro-/nanostructures and transferring them to flexible substrates," *Microelectron. Eng.*, vol. 98, pp. 502–507, 2012.

[49] D. Shakthivel, C. García Núñez, and R. Dahiya, "Inorganic semiconducting nanowires for flexible electronics," *United Scholars Publications, USA*, 2016.

[50] M.-C. Choi, Y. Kim, and C.-S. Ha, "Polymers for flexible displays: From material selection to device applications," *Prog. Polym. Sci.*, vol. 33, no. 6, pp. 581–630, 2008.

[51] M. Kwiat, S. Cohen, A. Pevzner, and F. Patolsky, "Large-scale ordered 1D-nanomaterials arrays: Assembly or not?," *Nano Today*, vol. 8, no. 6, pp. 677–694, 2013.

[52] Y.-Z. Long, M. Yu, B. Sun, C.-Z. Gu, and Z. Fan, "Recent advances in large-scale assembly of semiconducting inorganic nanowires and nanofibers for electronics, sensors and photovoltaics," *Chem. Soc. Rev.*, vol. 41, no. 12, pp. 4560–4580, 2012.

[53] A. Javey, S. Nam, R. S. Friedman, H. Yan, and C. M. Lieber, "Layer-by-layer assembly of nanowires for three-dimensional, multifunctional electronics," *Nano Lett.*, vol. 7, no. 3, pp. 773–777, 2007.

[54] D. Shakthivel, F. Liu, C. G. Núñez, W. Taube, and R. Dahiya, "Nanomaterials processing for flexible electronics," in *Industrial Electronics (ISIE), 2017 IEEE 26th International Symposium on*, 2017, pp. 2102–2106: IEEE.

[55] B. Su, Y. Wu, and L. Jiang, "The art of aligning one-dimensional (1D) nanostructures," *Chem. Soc. Rev.*, vol. 41, no. 23, pp. 7832–7856, 2012.

[56] R. Dahiya, G. Gottardi, and N. Laidani, "PDMS residues-free micro/macrostructures on flexible substrates," *Microelectron. Eng.*, vol. 136, pp. 57–62, 2015.

[57] Z. Fan, *et al.*, "Toward the development of printable nanowire electronics and sensors," *Adv. Mater.*, vol. 21, no. 37, pp. 3730–3743, 2009.

[58] N. Wang, Y. Cai, and R. Q. Zhang, "Growth of nanowires," *Mater. Sci. Eng. Rep.*, vol. 60, no. 1–6, pp. 1–51, 3/31/ 2008.

[59] A. Zhang, G. Zheng, and C. M. Lieber, *Nanowires: Building blocks for nanoscience and nanotechnology.* Springer, 2016.

[60] M. Amato, M. Palummo, R. Rurali, and S. Ossicini, "Silicon–germanium nanowires: chemistry and physics in play, from basic principles to advanced applications," *Chem. Rev.*, vol. 114, no. 2, pp. 1371–1412, 2013.

[61] N. Singh, et al., "Si, SiGe nanowire devices by top-down technology and their applications," *IEEE. T. Electron. Dev.*, vol. 55, no. 11, pp. 3107–3118, 2008.

[62] R. G. Hobbs, N. Petkov, and J. D. Holmes, "Semiconductor nanowire fabrication by bottom-up and top-down paradigms," *Chem. Mater.*, vol. 24, no. 11, pp. 1975–1991, 2012.

[63] R. S. Wagner and W. C. Ellis, "Vapor-liquid-solid mechanism of single crystal growth," *Appl. Phys. Lett*, vol. 4, no. 5, pp. 89–90, 1964.

[64] R. Wagner and C. Doherty, "Controlled vapor-liquid-solid growth of silicon crystals," *J. Electrochem. Soc.*, vol. 113, no. 12, pp. 1300–1305, 1966.

[65] R. Wagner and C. Ooherty, "Mechanism of branching and kinking during VLS crystal growth," *J. Electrochem. Soc.*, vol. 115, no. 1, pp. 93–99, 1968.

[66] X. Liu, Y.-Z. Long, L. Liao, X. Duan, and Z. Fan, "Large-scale integration of semiconductor nanowires for high-performance flexible electronics," *ACS Nano*, vol. 6, no. 3, pp. 1888–1900, 2012.

[67] D. Shakthivel, W. Taube, S. Raghavan, and R. Dahiya, "VLS growth mechanism of Si-nanowires for flexible electronics," in *IEEE 11th Conference on Ph. D. Research in Microelectronics and Electronics (PRIME)*, 2015, pp. 349–352.

[68] H. J. Fan, F. Bertram, A. Dadgar, J. Christen, A. Krost, and M. Zacharias, "Self-assembly of ZnO nanowires and the spatial resolved characterization of their luminescence," *Nanotechnology*, vol. 15, no. 11, p. 1401, 2004.

[69] Z. R. Dai, Z. W. Pan, and Z. L. Wang, "Novel nanostructures of functional oxides synthesized by thermal evaporation," *Adv. Funct. Mater*, vol. 13, no. 1, pp. 9–24, 2003.

[70] C. D. Scott, S. Arepalli, P. Nikolaev, and R. E. Smalley, "Growth mechanisms for single-wall carbon nanotubes in a laser-ablation process," *Appl. Phys. A-Mater.*, vol. 72, no. 5, pp. 573–580, May 2001.

[71] S. H. Jung, et al., "High-yield synthesis of multi-walled carbon nanotubes by arc discharge in liquid nitrogen," *Appl. Phys. A-Mater.*, vol. 76, no. 2, pp.285–286, Feb 2003.

[72] N. G. Shang, Y. Y. Tan, V. Stolojan, P. Papakonstantinou, and S. R. P. Silva, "High-rate low-temperature growth of vertically aligned carbon nanotubes," *Nanotechnology*, vol. 21, no. 50, p. 6, Dec 2010.

[73] Y. Y. Tan, *et al.*, "Photo-thermal chemical vapor deposition growth of graphene," *Carbon*, vol. 50, no. 2, pp.668–673, Feb 2012.

[74] J.-H. Ahn, *et al.*, "Heterogeneous three-dimensional electronics by use of printed semiconductor nanomaterials," *Science*, vol. 314, no. 5806, pp.1754–1757, 2006.

[75] R. Yerushalmi, Z. A. Jacobson, J. C. Ho, Z. Fan, and A. Javey, "Large scale, highly ordered assembly of nanowire parallel arrays by differential roll printing," *Appl. Phys. Lett*, vol. 91, no. 20, p. 203104, 2007.

[76] V. Schmidt, J. V. Wittemann, and U. Gösele, "Growth, thermodynamics, and electrical properties of silicon nanowires," *Chem. Rev.*, vol. 110, no. 1, pp. 361–388, 2010.

[77] V. Schmidt, J. V. Wittemann, S. Senz, and U. Gösele, "Silicon nanowires: a review on aspects of their growth and their electrical properties," *Adv. Mater.*, vol. 21, no. 25–26, pp. 2681–2702, 2009.

[78] B. O. Boskovic, V. Stolojan, R. U. A. Khan, S. Haq, and S. R. P. Silva, "Large-area synthesis of carbon nanofibres at room temperature," *Nat. Mater.*, vol. 1, no. 3, pp.165–168, Nov 2002.

[79] Y. Wang, V. Schmidt, S. Senz, and U. Gösele, "Epitaxial growth of silicon nanowires using an aluminium catalyst," *Nat. Nanotechnol.*, vol. 1, no. 3, pp.186–189, 2006.

[80] H. Schmid, *et al.*, "Patterned epitaxial vapor-liquid-solid growth of silicon nanowires on Si (111) using silane," *J. Appl. Phys.*, vol. 103, no. 2, p. 024304, 2008.

[81] K.-K. Lew and J. M. Redwing, "Growth characteristics of silicon nanowires synthesized by vapor–liquid–solid growth in nanoporous alumina templates," *J. Cryst. Growth.*, vol. 254, no. 1, pp.14–22, 2003.

[82] H. J. Fan, P. Werner, and M. Zacharias, "Semiconductor nanowires: from self-organization to patterned growth," *Small*, vol. 2, no. 6, pp.700–717, 2006.

[83] F. M. Ross, "Controlling nanowire structures through real time growth studies," *Rep. Prog. Phys.*, vol. 73, no. 11, p.114501, 2010.

[84] J. Hannon, S. Kodambaka, F. Ross, and R. Tromp, "The influence of the surface migration of gold on the growth of silicon nanowires," *Nature*, vol. 440, no. 7080, pp.69–71, 2006.

[85] C.-Y. Wen, *et al.*, "Periodically changing morphology of the growth interface in Si, Ge, and GaP nanowires," *Phys. Rev. Lett.*, vol. 107, no. 2, p. 025503, 2011.

[86] O. Moutanabbir, D. Isheim, H. Blumtritt, S. Senz, E. Pippel, and D. N. Seidman, "Colossal injection of catalyst atoms into silicon nanowires," *Nature*, vol. 496, no. 7443, pp.78–82, 2013.

[87] J. E. Allen, *et al.*, "High-resolution detection of Au catalyst atoms in Si nanowires," *Nat. Nanotechnol.*, vol. 3, no. 3, pp.168–173, 2008.

[88] J. L. Lensch-Falk, E. R. Hemesath, D. E. Perea, and L. J. Lauhon, "Alternative catalysts for VSS growth of silicon and germanium nanowires," *J. Mater. Chem.*, 10.1039/B817391E vol. 19, no. 7, pp.849–857, 2009.

[89] T. Baron, M. Gordon, F. Dhalluin, C. Ternon, P. Ferret, and P. Gentile, "Si nanowire growth and characterization using a microelectronics-compatible catalyst: PtSi," *Appl. Phys. Lett.*, vol. 89, no. 23, p.233111, 2006.

[90] L. Cao, B. Garipcan, J. S. Atchison, C. Ni, B. Nabet, and J. E. Spanier, "Instability and transport of metal catalyst in the growth of tapered silicon nanowires," *Nano Lett.*, vol. 6, no. 9, pp. 1852–1857, 2006.

[91] V. Schmidt, S. Senz, and U. Gösele, "Diameter dependence of the growth velocity of silicon nanowires synthesized via the vapor-liquid-solid mechanism," *Phys. Rev. B*, vol. 75, no. 4, p.045335, 2007.

[92] C. W. Pinion, D. P. Nenon, J. D. Christesen, and J. F. Cahoon, "Identifying crystallization- and incorporation-limited regimes during vapor–liquid–solid growth of Si nanowires," *ACS Nano*, vol. 8, no. 6, pp. 6081–6088, 2014.

[93] V. G. Dubrovskii, *Nucleation theory and growth of nanostructures*. Springer, 2014.

[94] A. Handbook, "Vol. 3: Alloy phase diagrams," *ASM International*, vol. 9, p. 2, 1992.

[95] M. K. Sunkara, S. Sharma, R. Miranda, G. Lian, and E. Dickey, "Bulk synthesis of silicon nanowires using a low-temperature vapor–liquid–solid method," *Appl. Phys. Lett*, vol. 79, no. 10, pp. 1546–1548, 2001.

[96] S. Sharma and M. Sunkara, "Direct synthesis of single-crystalline silicon nanowires using molten gallium and silane plasma," *Nanotechnology*, vol. 15, no. 1, p. 130, 2003.

[97] S.-Y. Choi, W. Y. Fung, and W. Lu, "Growth and electrical properties of Al-catalyzed Si nanowires," *Appl. Phys. Lett.*, vol. 98, no. 3, p. 033108, 2011.

[98] J. Arbiol, B. Kalache, P. R. i. Cabarrocas, J. R. Morante, and A. F. i. Morral, "Influence of Cu as a catalyst on the properties of silicon nanowires synthesized by the vapour–solid–solid mechanism," *Nanotechnology*, vol. 18, no. 30, p. 305606, 2007.

[99] S. Khan, L. Lorenzelli, and R. S. Dahiya, "Technologies for printing sensors and electronics over large flexible substrates: a review," *IEEE. Sens. J.*, vol. 15, no. 6, pp. 3164–3185, 2015.

[100] D.-H. Kim, et al., "Stretchable and foldable silicon integrated circuits," *Science*, vol. 320, no. 5875, pp. 507–511, 2008.

[101] S. J. Kang, et al., "High-performance electronics using dense, perfectly aligned arrays of single-walled carbon nanotubes," *Nat. Nanotechnol.*, vol. 2, no. 4, p. 230, 2007.

[102] H. Han, Z. Huang, and W. Lee, "Metal-assisted chemical etching of silicon and nanotechnology applications," *Nano Today*, vol. 9, no. 3, pp. 271–304, 2014.

[103] W. Lu and C. M. Lieber, "Nanoelectronics from the bottom up," *Nat. Mater.*, vol. 6, no. 11, pp. 841–850, 2007.

[104] Y. Cui and C. M. Lieber, "Functional nanoscale electronic devices assembled using silicon nanowire building blocks," *Science*, vol. 291, no. 5505, pp. 851–853, 2001.

[105] B. Fuhrmann, H. S. Leipner, H.-R. Höche, L. Schubert, P. Werner, and U. Gösele, "Ordered arrays of silicon nanowires produced by nanosphere lithography and molecular beam epitaxy," *Nano Lett.*, vol. 5, no. 12, pp. 2524–2527, 2005.

[106] A. M. Morales and C. M. Lieber, "A laser ablation method for the synthesis of crystalline semiconductor nanowires," *Science*, vol. 279, no. 5348, pp. 208–211, 1998.

[107] Y. Zhang, et al., "Silicon nanowires prepared by laser ablation at high temperature," *Appl. Phys. Lett.*, vol. 72, no. 15, pp. 1835–1837, 1998.

[108] J.-S. Lee, M.-I. Kang, S. Kim, M.-S. Lee, and Y.-K. Lee, "Growth of zinc oxide nanowires by thermal evaporation on vicinal Si (100) substrate," *J. Cryst. Growth.*, vol. 249, no. 1, pp. 201–207, 2003.

[109] M. J. Bierman, Y. A. Lau, A. V. Kvit, A. L. Schmitt, and S. Jin, "Dislocation-driven nanowire growth and Eshelby twist," *Science*, vol. 320, no. 5879, pp. 1060–1063, 2008.

[110] M. M. Khalaf, H. G. Ibrahimov, and E. H. Ismailov, "Nanostructured materials: importance, synthesis and characterization—a review," *Chemistry Journal* vol. 2, no. 3, pp. 118–125, 2012.

[111] M. L. Hitchman and K. F. Jensen, *Chemical vapor deposition: principles and applications.* Elsevier, 1993.

[112] H. O. Pierson, *Handbook of chemical vapor deposition: principles, technology and applications.* William Andrew, 1999.

[113] B. Kim, J. Tersoff, S. Kodambaka, M. Reuter, E. Stach, and F. Ross, "Kinetics of individual nucleation events observed in nanoscale vapor-liquid-solid growth," *Science*, vol. 322, no. 5904, pp. 1070–1073, 2008.

[114] S. Hofmann, *et al.*, "In situ observations of catalyst dynamics during surface-bound carbon nanotube nucleation," *Nano Lett.*, vol. 7, no. 3, pp. 602–608, 2007.

[115] A. Gamalski, C. Ducati, and S. Hofmann, "Cyclic supersaturation and triple phase boundary dynamics in germanium nanowire growth," *J. Phys. Chem. B*, vol. 115, no. 11, pp. 4413–4417, 2011.

[116] J. Xiang, W. Lu, Y. Hu, Y. Wu, H. Yan, and C. M. Lieber, "Ge/Si nanowire heterostructures as high-performance field-effect transistors," *Nature*, vol. 441, no. 7092, pp. 489–493, 2006.

[117] F. Qian, Y. Li, S. Gradecak, D. Wang, C. J. Barrelet, and C. M. Lieber, "Gallium nitride-based nanowire radial heterostructures for nanophotonics," *Nano Lett.*, vol. 4, no. 10, pp. 1975–1979, 2004.

[118] H. Schmid, M. T. Björk, J. Knoch, S. Karg, H. Riel, and W. Riess, "Doping limits of grown in situ doped silicon nanowires using phosphine," *Nano Lett.*, vol. 9, no. 1, pp. 173–177, 2008.

[119] J. Wallentin and M. T. Borgström, "Doping of semiconductor nanowires," *J. Mater. Res.*, vol. 26, no. 17, pp. 2142–2156, 2011.

[120] C. Yang, Z. Zhong, and C. M. Lieber, "Encoding electronic properties by synthesis of axial modulation-doped silicon nanowires," *Science*, vol. 310, no. 5752, pp. 1304–1307, 2005.

[121] E. Givargizov, "Fundamental aspects of VLS growth," *J. Cryst. Growth.*, vol. 31, pp. 20–30, 1975.

[122] E. I. Givargizov and N. N. Sheftal, "Morphology of silicon whiskers grown by the VLS-technique," *J. Cryst. Growth*, vol. 9, no. 0, pp. 326–329, 5, 1971.

[123] G. Bootsma and H. Gassen, "A quantitative study on the growth of silicon whiskers from silane and germanium whiskers from germane," *J. Cryst. Growth*, vol. 10, no. 3, pp. 223–234, 1971.

[124] S. Kodambaka, J. Tersoff, M. Reuter, and F. Ross, "Diameter-independent kinetics in the vapor-liquid-solid growth of Si nanowires," *Phys. Rev. Lett.*, vol. 96, no. 9, p. 096105, 2006.

[125] K. J. Laidler, *Chemical Kinetics*. Delhi: Pearson Education, 2008.

[126] L. Fröberg, W. Seifert, and J. Johansson, "Diameter-dependent growth rate of InAs nanowires," *Phys. Rev. B*, vol. 76, no. 15, p. 153401, 2007.

[127] D. Kashchiev, "Dependence of the growth rate of nanowires on the nanowire diameter," *Cryst. Growth. Des.*, vol. 6, no. 5, pp. 1154–1156, 2006.

[128] D. Shakthivel and S. Raghavan, "Vapor-liquid-solid growth of Si nanowires: A kinetic analysis," *J. Appl. Phys.*, vol. 112, no. 2, p. 024317, 2012.

[129] T. Mårtensson, *et al.*, "Epitaxial III–V Nanowires on Silicon," *Nano Lett.*, vol. 4, no. 10, pp. 1987–1990, 2004.

[130] R. Q. Zhang, Y. Lifshitz, and S. T. Lee, "Oxide-assisted growth of semiconducting nanowires," *Adv. Mater.*, vol. 15, no. 7–8, pp. 635–640, 2003.

[131] F. Kolb, *et al.*, "Analysis of silicon nanowires grown by combining SiO evaporation with the VLS mechanism," *J. Electrochem. Soc.*, vol. 151, no. 7, pp. G472–G475, 2004.

[132] M. H. Huang, Y. Wu, H. Feick, N. Tran, E. Weber, and P. Yang, "Catalytic growth of zinc oxide nanowires by vapor transport," *Adv. Mater.*, vol. 13, no. 2, pp. 113–116, 2001.

[133] A. I. Persson, M. W. Larsson, S. Stenström, B. J. Ohlsson, L. Samuelson, and L. R. Wallenberg, "Solid-phase diffusion mechanism for GaAs nanowire growth," *Nat. Mater.*, vol. 3, no. 10, pp. 677–681, 2004.

[134] T. Kamins, X. Li, R. S. Williams, and X. Liu, "Growth and structure of chemically vapor deposited Ge nanowires on Si substrates," *Nano Lett.*, vol. 4, no. 3, pp. 503–506, 2004.

[135] V. Dubrovskii, N. Sibirev, and G. Cirlin, "Kinetic model of the growth of nanodimensional whiskers by the vapor-liquid-crystal mechanism," *Tech. Phys. Lett.*, vol. 30, no. 8, pp. 682–686, 2004.

[136] S. Hofmann, *et al.*, "Ledge-flow-controlled catalyst interface dynamics during Si nanowire growth," *Nat. Mater.*, vol. 7, no. 5, p. 372, 2008.

[137] H. Zhao, S. Zhou, Z. Hasanali, and D. Wang, "Influence of pressure on silicon nanowire growth kinetics," *J. Phys. Chem. B*, vol. 112, no. 15, pp. 5695–5698, 2008.

[138] D. Pal, M. Kowar, A. Daw, and P. Roy, "Modelling of silicon epitaxy using silicon tetrachloride as the source," *Microelectr. J.*, vol. 26, no. 6, pp. 507–514, 1995.

[139] L. Zambov, "Kinetics of homogeneous decomposition of silane," *J. Cryst. Growth*, vol. 125, no. 1, pp. 164–174, 1992.

[140] V. Dubrovskii, N. Sibirev, G. Cirlin, J. Harmand, and V. Ustinov, "Theoretical analysis of the vapor-liquid-solid mechanism of nanowire growth during molecular beam epitaxy," *Phys. Rev. E*, vol. 73, no. 2, p. 021603, 2006.

[141] J. Johansson, B. A. Wacaser, K. A. Dick, and W. Seifert, "Growth related aspects of epitaxial nanowires," *Nanotechnology*, vol. 17, no. 11, p. S355, 2006.

[142] J. P. Hirth and G. M. Pound, *Condensation and evaporation; nucleation and growth kinetics*. Macmillan, 1963.

[143] I. Markov, "Crystal growth for beginners: fundamentals of nucleation," *Crystal Growth and Epitaxy*, p. 69, 1995.

[144] B. A. Wacaser, K. A. Dick, J. Johansson, M. T. Borgström, K. Deppert, and L. Samuelson, "Preferential interface nucleation: an expansion of the VLS growth mechanism for nanowires," *Adv. Mater.*, vol. 21, no. 2, pp. 153–165, 2009.

[145] D. Shakthivel, S. Rathkanthiwar, and S. Raghavan, "Si nanowire growth on sapphire: Classical incubation, reverse reaction, and steady state supersaturation," *J. Appl. Phys.*, vol. 117, no. 16, p. 164302, 2015.

[146] B. Kalache, P. R. i Cabarrocas, and A. F. i Morral, "Observation of incubation times in the nucleation of silicon nanowires obtained by the vapor–liquid–solid method," *Jpn. J. Appl. Phys.*, vol. 45, no. 2L, p. L190, 2006.

[147] F. Kreupl, *et al.*, "Carbon nanotubes in interconnect applications," *Microelectron. Eng.*, vol. 64, no. 1–4, pp. 399–408, Oct 2002.

[148] H. W. Kroto, J. R. Heath, S. C. Obrien, R. F. Curl, and R. E. Smalley, "C-60 – Buckminsterfullerene," (in English), *Nature*, vol. 318, no. 6042, pp. 162–163, 1985.

[149] S. Iijima, "Helical Microtubules of Graphitic Carbon," *Nature*, vol. 354, no. 6348, pp. 56–58, Nov 1991.

[150] B. Q. Wei, R. Vajtai, and P. M. Ajayan, "Reliability and current carrying capacity of carbon nanotubes," (in English), *Appl. Phys. Lett.*, vol. 79, no. 8, pp. 1172–1174, Aug 2001.

[151] S. Frank, P. Poncharal, Z. L. Wang, and W. A. de Heer, "Carbon nanotube quantum resistors" (in English), *Science*, vol. 280, no. 5370, pp. 1744–1746, Jun 1998.

[152] M. S. Dresselhaus, G. Dresselhaus, and P. Avouris, *Carbon nanotubes: synthesis, structure, properties, and applications*, Springer Books 2001.

[153] K. Hata, D. N. Futaba, K. Mizuno, T. Namai, M. Yumura, and S. Iijima, "Water-assisted highly efficient synthesis of impurity-free single-waited carbon nanotubes," *Science*, vol. 306, no. 5700, pp. 1362–1364, Nov 19 2004.

[154] E. T. Thostenson, Z. F. Ren, and T. W. Chou, "Advances in the science and technology of carbon nanotubes and their composites: a review," *Compos. Sci. Technol.*, vol. 61, no. 13, pp. 1899–1912, 2001.

[155] M. S. Dresselhaus, G. Dresselhaus, R. Saito, and A. Jorio, "Raman spectroscopy of carbon nanotubes," *Phys. Rep.*, vol. 409, no. 2, pp. 47–99, Mar 2005.

[156] R. Saito, G. Dresselhaus, and M. S. Dresselhaus, *Physical properties of carbon nanotubes*, London: Imperial College Press, 1998.

[157] V. N. Popov, "Carbon nanotubes: properties and application," *Mater. Sci. Eng. Rep.*, vol. 43, no. 3, pp. 61–102, Jan 15 2004.

[158] M. S. Dresselhaus, A. Jorio, and R. Saito, "Characterizing graphene, graphite, and carbon nanotubes by raman spectroscopy," in *Annu. Rev. Conden. Ma. P*, vol. 1, 2010, pp. 89–108.

[159] S. Park, M. Vosguerichian, and Z. Bao, "A review of fabrication and applications of carbon nanotube film-based flexible electronics," *Nanoscale*, vol. 5, no. 5, pp. 1727–1752, 2013.

[160] S. J. Tans, *et al.*, "Individual single-wall carbon nanotubes as quantum wires," *Nature*, vol. 386, no. 6624, pp. 474–477, Apr 1997.

[161] A. Bachtold, *et al.*, "Aharonov-Bohm oscillations in carbon nanotubes," *Nature*, vol. 397, no. 6721, pp. 673–675, 1999.

[162] S. Roth, V. Krstic, and G. Rikken, "Quantum transport in carbon nanotubes," *Curr. Appl. Phys.*, vol. 2, no. 2, pp. 155–161, 2002.

[163] R. Maruyama, Y. W. Nam, J. H. Han, and M. S. Strano, "Well-defined single-walled carbon nanotube fibers as quantum wires: Ballistic conduction over micrometer-length scales," *Curr. Appl. Phys.*, vol. 11, no. 6, pp. 1414–1418, Nov 2011.

[164] J. Kong, *et al.*, "Quantum interference and ballistic transmission in nanotube electron waveguides," *Phys. Rev. Lett.*, vol. 87, no. 10, p. 4, Sep 2001, Art. no. 106801.

[165] A. Javey, J. Guo, Q. Wang, M. Lundstrom, and H. J. Dai, "Ballistic carbon nanotube field-effect transistors," *Nature*, vol. 424, no. 6949, pp. 654–657, Aug 2003.

[166] C. Rutherglen and P. Burke, "Nanoelectromagnetics: circuit and electromagnetic properties of carbon nanotubes," *Small*, vol. 5, no. 8, pp. 884–906, Apr 2009.

[167] Y. W. Fan, B. R. Goldsmith, and P. G. Collins, "Identifying and counting point defects in carbon nanotubes," *Nat. Mater.*, vol. 4, no. 12, pp. 906–911, Dec 2005.

[168] Y. Ando, X. Zhao, T. Sugai, and M. Kumar, "Growing carbon nanotubes," *Mater. Today.*, vol. 7, no. 10, pp. 22–29, 2004.

[169] Y. Ando, "Carbon nanotube: the inside story," *J. Nanosci. Nanotechno.*, vol. 10, no. 6, pp. 3726–3738, Jun 2010.

[170] D. T. Colbert, *et al.*, "Growth and sintering of fullerene nanotubes," *Science*, vol. 266, no. 5188, pp. 1218–1222, Nov 1994.

[171] T. W. Ebbesen and P. M. Ajayan, "Large-scale synthesis of carbon nanotubes," *Nature* vol. 358, pp. 220–222, 1992.

[172] T. W. Ebbesen, P. M. Ajayan, H. Hiura, and K. Tanigaki, "Purification of nanotubes," *Nature*, Letter vol. 367, no. 6463, pp. 519–519, Feb 1994.

[173] J.-P. Tessonnier and D. S. Su, "Recent progress on the growth mechanism of carbon nanotubes: a review," *Chemsuschem*, vol. 4, no. 7, pp. 824–847, 2011.

[174] K. Anazawa, K. Shimotani, C. Manabe, H. Watanabe, and M. Shimizu, "High-purity carbon nanotubes synthesis method by an arc discharging in magnetic field," *Appl. Phys. Lett.*, vol. 81, no. 4, pp. 739–741, Jul 2002.

[175] M. C. McAlpine, H. Ahmad, D. Wang, and J. R. Heath, "Highly ordered nanowire arrays on plastic substrates for ultrasensitive flexible chemical sensors," *Nat. Mater.*, vol. 6, no. 5, p. 379, 2007.

[176] S. Farhat, *et al.*, "Diameter control of single-walled carbon nanotubes using argon-helium mixture gases," *J. Chem. Phys.*, vol. 115, no. 14, pp. 6752–6759, Oct 2001.

[177] T. Guo, *et al.*, "Uranium stabilization of C_{28} – a tetravalent fullerene," *Science*, vol. 257, no. 5077, pp. 1661–1664, Sep 1992.

[178] A. Thess, *et al.*, "Crystalline ropes of metallic carbon nanotubes," *Science*, vol. 273, no. 5274, pp. 483–487, Jul 1996.

[179] S. Bandow, A. M. Rao, K. A. Williams, A. Thess, R. E. Smalley, and P. C. Eklund, "Purification of single-wall carbon nanotubes by microfiltration," *J. Phys. Chem. B*, Letter vol. 101, no. 44, pp. 8839–8842, Oct 1997.

[180] H. Ishii, *et al.*, "Direct observation of Tomonaga-Luttinger-liquid state in carbon nanotubes at low temperatures," *Nature*, vol. 426, no. 6966, pp. 540–544, Dec 2003.

[181] P. C. Eklund, *et al.*, "Large-scale production of single-walled carbon nanotubes using ultrafast pulses from a free electron laser," *Nano Lett.*, vol. 2, no. 6, pp. 561–566, Jun 2002.

[182] A. P. Bolshakov, *et al.*, "A novel CW laser-powder method of carbon single-wall nanotubes production," *Diam. Relat. Mater.*, vol. 11, no. 3–6, pp. 927–930, Mar–Jun 2002.

[183] M. Endo, K. Takeuchi, S. Igarashi, K. Kobori, M. Shiraishi, and H. W. Kroto, "The production and structure of pyrolytic carbon nanotubes (PNTs)," *J. Phys. Chem. Solids.*, vol. 54, no. 12, pp. 1841–1848, Dec 1993.

[184] M. Kumar and Y. Ando, "Chemical vapor deposition of carbon nanotubes: a review on growth mechanism and mass production," *J. Nanosci. Nanotechno.*, vol. 10, no. 6, pp. 3739–3758, Jun 2010.

[185] Z. F. Ren, *et al.*, "Growth of a single freestanding multiwall carbon nanotube on each nanonickel dot," *Appl. Phys. Lett.*, vol. 75, no. 8, pp. 1086–1088, Aug 1999.

[186] Z. F. Ren, et al., "Synthesis of large arrays of well-aligned carbon nanotubes on glass," Science, vol. 282, no. 5391, pp. 1105–1107, Nov 1998.

[187] M. Yudasaka, R. Kikuchi, T. Matsui, Y. Ohki, S. Yoshimura, and E. Ota, "Specific conditions for Ni catalyzed carbon nanotube growth by chemical-vapor-deposition," Appl. Phys. Lett., vol. 67, no. 17, pp. 2477–2479, Oct 1995.

[188] M. Yudasaka, R. Kikuchi, Y. Ohki, E. Ota, and S. Yoshimura, "Behavior of Ni in carbon nanotube nucleation," Appl. Phys. Lett., vol. 70, no. 14, pp. 1817–1818, Apr 1997.

[189] C.-M. Seah, S.-P. Chai, and A. R. Mohamed, "Synthesis of aligned carbon nanotubes," Carbon, vol. 49, no. 14, pp. 4613–4635, Nov 2011.

[190] M. A. Azam, N. S. A. Manaf, E. Talib, and M. S. A. Bistamam, "Aligned carbon nanotube from catalytic chemical vapor deposition technique for energy storage device: a review," Ionics, vol. 19, no. 11, pp. 1455–1476, 2013.

[191] S. B. Sinnott, et al., "Model of carbon nanotube growth through chemical vapor deposition," Chem. Phys. Lett., vol. 315, no. 1–2, pp. 25–30, Dec 1999.

[192] Q. Li, et al., "Sustained growth of ultralong carbon nanotube arrays for fiber spinning," Adv. Mater., vol. 18, no. 23, pp. 3160–3163, Dec 4 2006.

[193] P. B. Amama, et al., "Role of water in super growth of single-walled carbon nanotube carpets," Nano Lett., vol. 9, no. 1, pp. 44–49, 2009.

[194] Q. Wen, et al., "Growing 20 cm long DWNTs/TWNTs at a rapid growth rate of 80–90 um/s," Chem. Mater., vol. 22, no. 4, pp. 1294–1296, 2010.

[195] A. M. Cassell, J. A. Raymakers, J. Kong, and H. J. Dai, "Large scale CVD synthesis of single-walled carbon nanotubes," J. Phys. Chem. B, vol. 103, no. 31, pp. 6484–6492, Aug 1999.

[196] H. J. Dai, et al., "Controlled chemical routes to nanotube architectures, physics, and devices," (in English), J. Phys. Chem. B, vol. 103, no. 51, pp. 11246–11255, Dec 1999.

[197] J. Kong, A. M. Cassell, and H. J. Dai, "Chemical vapor deposition of methane for single-walled carbon nanotubes," Chem. Phys. Lett., vol. 292, no. 4–6, pp. 567–574, Aug 1998.

[198] J. Kong, H. T. Soh, A. M. Cassell, C. F. Quate, and H. J. Dai, "Synthesis of individual single-walled carbon nanotubes on patterned silicon wafers," Nature, vol. 395, no. 6705, pp. 878–881, Oct 1998.

[199] M. Ahmad, J. V. Anguita, V. Stolojan, J. D. Carey, and S. R. P. Silva, "Efficient coupling of optical energy for rapid catalyzed nanomaterial growth: high-quality carbon nanotube synthesis at low substrate

temperatures," *ACS Appl. Mater. Inter*, vol. 5, no. 9, pp. 3861–3866, May 8 2013.

[200] M. Ahmad, *et al.*, "High quality carbon nanotubes on conductive substrates grown at low temperatures," *Adv. Funct. Mater*, vol. 25, no. 28, pp. 4419–4429, 2015.

[201] M. Ahmad, "Carbon nanotube based integrated circuit interconnects," University of Surrey, Faculty of Engineering and Physical Sciences, Department of Electronic Engineering Thesis (Ph.D.) – University of Surrey, 2013.

[202] G. Y. Chen, B. Jensen, V. Stolojan, and S. R. P. Silva, "Growth of carbon nanotubes at temperatures compatible with integrated circuit technologies," *Carbon*, vol. 49, no. 1, pp. 280–285, Jan 2011.

[203] M. Su, B. Zheng, and J. Liu, "A scalable CVD method for the synthesis of single-walled carbon nanotubes with high catalyst productivity," *Chem. Phys. Lett.*, vol. 322, no. 5, pp. 321–326, May 2000.

[204] R. Alexandrescu, *et al.*, "Synthesis of carbon nanotubes by CO_2-laser-assisted chemical vapour deposition," *Infrared. Phys. Techn.*, vol. 44, no. 1, pp. 43–50, Feb 2003.

[205] S. Maruyama, Y. Miyauchi, Y. Murakami, and S. Chiashi, "Optical characterization of single-walled carbon nanotubes synthesized by catalytic decomposition of alcohol," *New. J. Phys.*, vol. 5, p. 12, Oct 2003.

[206] C. Bower, W. Zhu, S. H. Jin, and O. Zhou, "Plasma-induced alignment of carbon nanotubes," *Appl. Phys. Lett.*, vol. 77, no. 6, pp. 830–832, Aug 2000.

[207] V. I. Merkulov, D. H. Lowndes, Y. Y. Wei, G. Eres, and E. Voelkl, "Patterned growth of individual and multiple vertically aligned carbon nanofibers," *Appl. Phys. Lett.*, vol. 76, no. 24, pp. 3555–3557, Jun 2000.

[208] K. B. K. Teo, *et al.*, "Uniform patterned growth of carbon nanotubes without surface carbon," *Appl. Phys. Lett*, vol. 79, no. 10, pp. 1534–1536, Sep 2001.

[209] J. V. Anguita, D. C. Cox, M. Ahmad, Y. Y. Tan, J. Allam, and S. R. P. Silva, "Highly transmissive carbon nanotube forests grown at low substrate temperature," *Adv. Funct. Mater.*, 2013.

[210] M. Chhowalla, *et al.*, "Growth process conditions of vertically aligned carbon nanotubes using plasma enhanced chemical vapor deposition," *J. Appl. Phys.*, vol. 90, no. 10, pp. 5308–5317, Nov 15 2001.

[211] S. Maruyama, R. Kojima, Y. Miyauchi, S. Chiashi, and M. Kohno, "Low-temperature synthesis of high-purity single-walled carbon nanotubes from alcohol," *Chem. Phys. Lett.*, vol. 360, no. 3–4, pp. 229–234, Jul 10 2002.

[212] R. Sen, A. Govindaraj, and C. N. R. Rao, "Carbon nanotubes by the metallocene route," *Chem. Phys. Lett.*, vol. 267, no. 3–4, pp. 276–280, Mar 21 1997.

[213] M. Kumar and Y. Ando, "A simple method of producing aligned carbon nanotubes from an unconventional precursor – Camphor," *Chem. Phys. Lett.*, vol. 374, no. 5–6, pp. 521–526, Jun 18 2003.

[214] Q. W. Li, H. Yan, J. Zhang, and Z. F. Liu, "Effect of hydrocarbons precursors on the formation of carbon nanotubes in chemical vapor deposition," *Carbon*, vol. 42, no. 4, pp. 829–835, 2004.

[215] D. Yuan, L. Ding, H. Chu, Y. Feng, T. P. McNicholas, and J. Liu, "Horizontally aligned single-walled carbon nanotube on quartz from a large variety of metal catalysts," *Nano Lett.*, vol. 8, no. 8, pp. 2576–2579, Aug 2008.

[216] R. Seidel, G. S. Duesberg, E. Unger, A. P. Graham, M. Liebau, and F. Kreupl, "Chemical vapor deposition growth of single-walled carbon nanotubes at 600 degrees C and a simple growth model," *J. Phys. Chem. B*, vol. 108, no. 6, pp. 1888–1893, Feb 12 2004.

[217] E. R. Meshot, D. L. Plata, S. Tawfick, Y. Zhang, E. A. Verploegen, and A. J. Hart, "Engineering vertically aligned carbon nanotube growth by decoupled thermal treatment of precursor and catalyst," *ACS Nano*, vol. 3, no. 9, pp. 2477–2486, Sep 2009.

[218] K. Y. Lee, et al., "Vertically aligned growth of carbon nanotubes with long length and high density," *J. Vac. Sci. Technol. B.*, vol. 23, no. 4, pp. 1450–1453, Jul–Aug 2005.

[219] G. D. Nessim, et al., "Tuning of vertically-aligned carbon nanotube diameter and areal density through catalyst pre-treatment," *Nano Lett.*, vol. 8, no. 11, pp. 3587–3593, Nov 2008.

[220] J. D. Carey, L. L. Ong, and S. R. P. Silva, "Formation of low-temperature self-organized nanoscale nickel metal islands," *Nanotechnology*, vol. 14, no. 11, pp. 1223–1227, Nov 2003.

[221] B. C. Bayer, et al., "Co-catalytic solid-state reduction applied to carbon nanotube growth," *J. Phys. Chem. B*, vol. 116, no. 1, pp. 1107–1113, Jan 12 2012.

[222] H. Ago, K. Nakamura, N. Uehara, and M. Tsuji, "Roles of metal-support interaction in growth of single- and double-walled carbon nanotubes studied with diameter-controlled iron particles supported on MgO," *J. Phys. Chem. B*, vol. 108, no. 49, pp. 18908–18915, Dec 9 2004.

[223] Y. J. Jung, B. Q. Wei, R. Vajtai, and P. M. Ajayan, "Mechanism of selective growth of carbon nanotubes on SiO$_2$/Si patterns," *Nano Lett.*, vol. 3, no. 4, pp. 561–564, Apr 2003.

[224] S. Esconjauregui, *et al.*, "Growth of ultrahigh density vertically aligned carbon nanotube forests for interconnects," *ACS Nano*, vol. 4, no. 12, pp. 7431–7436, Dec 2010.

[225] D. Yokoyama, *et al.*, "Low temperature grown carbon nanotube interconnects using inner shells by chemical mechanical polishing," *Appl. Phys. Lett.*, vol. 91, no. 26, p. 263101, Dec 24 2007.

[226] H. Okuno, *et al.*, "CNT integration on different materials suitable for VLSI interconnects," *C. R. Phys.*, vol. 11, no. 5–6, pp. 381–388, Jun–Jul 2010.

[227] M. H. van der Veen, *et al.*, "Electrical characterization of CNT contacts with Cu Damascene top contact," *Microelectron. Eng.*, vol. 106, pp. 106–111, Jun 2013.

[228] J. Li, *et al.*, "Bottom-up approach for carbon nanotube interconnects," *Appl. Phys. Lett*, vol. 82, no. 15, pp. 2491–2493, Apr 2003.

[229] H. Zhang, B. Wu, W. Hu, and Y. Liu, "Separation and/or selective enrichment of single-walled carbon nanotubes based on their electronic properties," *Chem. Soc. Rev.*, vol. 40, no. 3, pp. 1324–1336, 2011.

[230] A. R. Harutyunyan, *et al.*, "Preferential growth of single-walled carbon nanotubes with metallic conductivity," *Science*, vol. 326, no. 5949, p. 116, 2009.

[231] R. Krupke, F. Hennrich, H. v. Löhneysen, and M. M. Kappes, "Separation of metallic from semiconducting single-walled carbon nanotubes," *Science*, vol. 301, no. 5631, p. 344, 2003.

[232] N. Mureau, E. Mendoza, S. R. P. Silva, K. F. Hoettges, and M. P. Hughes, "In situ and real time determination of metallic and semiconducting single-walled carbon nanotubes in suspension via dielectrophoresis," *Appl. Phys. Lett*, vol. 88, no. 24, p. 243109, 2006.

[233] P. G. Collins, M. S. Arnold, and P. Avouris, "Engineering carbon nanotubes and nanotube circuits using electrical breakdown," *Science*, vol. 292, no. 5517, p. 706, 2001.

[234] M. Yudasaka, M. Zhang, and S. Iijima, "Diameter-selective removal of single-wall carbon nanotubes through light-assisted oxidation," *Chem. Phys. Lett.*, vol. 374, no. 1, pp. 132–136, 2003.

[235] M. Zheng, *et al.*, "Structure-based carbon nanotube sorting by sequence-dependent DNA assembly," *Science*, vol. 302, no. 5650, p. 1545, 2003.

[236] X. Tu, S. Manohar, A. Jagota, and M. Zheng, "DNA sequence motifs for structure-specific recognition and separation of carbon nanotubes," *Nature*, vol. 460, p. 250, 2009.

[237] T. Takeshi, J. Hehua, M. Yasumitsu, and K. Hiromichi, "High-yield separation of metallic and semiconducting single-wall carbon nanotubes by agarose gel electrophoresis," *Appl. Phys. Express*, vol. 1, no. 11, p. 114001, 2008.

[238] T. Tanaka, *et al.*, "Simple and scalable gel-based separation of metallic and semiconducting carbon nanotubes," *Nano Lett.*, vol. 9, no. 4, pp. 1497–1500, 2009.

[239] W. H. Duan, Q. Wang, and F. Collins, "Dispersion of carbon nanotubes with SDS surfactants: a study from a binding energy perspective," *Chem. Sci.*, vol. 2, no. 7, pp. 1407–1413, 2011.

[240] C. A. Silvera-Batista, D. C. Scott, S. M. McLeod, and K. J. Ziegler, "A mechanistic study of the selective retention of SDS-suspended single-wall carbon nanotubes on agarose gels," *J. Phys. Chem. B*, vol. 115, no. 19, pp. 9361–9369, 2011.

[241] M. S. Arnold, A. A. Green, J. F. Hulvat, S. I. Stupp, and M. C. Hersam, "Sorting carbon nanotubes by electronic structure using density differentiation," *Nat. Nanotechnol.*, vol. 1, p. 60, 2006.

[242] H. Liu, D. Nishide, T. Tanaka, and H. Kataura, "Large-scale single-chirality separation of single-wall carbon nanotubes by simple gel chromatography," *Nat. Commun.*, vol. 2, p. 309, 2011.

[243] S. Xu and Z. L. Wang, "One-dimensional ZnO nanostructures: solution growth and functional properties," *Nano. Res.*, vol. 4, no. 11, pp. 1013–1098, November 01 2011.

[244] Z. L. Wang, "ZnO nanowire and nanobelt platform for nanotechnology," *Mater. Sci. Eng. Rep.*, vol. 64, no. 3, pp. 33–71, 2009.

[245] S. Das and S. Ghosh, "Fabrication of different morphologies of ZnO superstructures in presence of synthesized ethylammonium nitrate (EAN) ionic liquid: synthesis, characterization and analysis," *Dalton. T.*, vol. 42, no. 5, pp. 1645–1656, 2013.

[246] Z. W. Pan, Z. R. Dai, and Z. L. Wang, "Nanobelts of semiconducting oxides," *Science*, vol. 291, no. 5510, p. 1947, 2001.

[247] W. I. Park, G. C. Yi, M. Y. Kim, and S. J. Pennycook, "ZnO nanoneedles grown vertically on Si substrates by non-catalytic vapor-phase epitaxy," *Adv. Mater.*, vol. 14, no. 24, pp. 1841–1843, Dec 2002.

[248] Y. W. Heo, *et al.*, "Site-specific growth of ZnO nanorods using catalysis-driven molecular-beam epitaxy," *Appl. Phys. Lett.*, vol. 81, no. 16, pp. 3046–3048, 2002.

[249] J. I. Hong, J. Bae, Z. L. Wang, and R. L. Snyder, "Room-temperature, texture-controlled growth of ZnO thin films and their application for growing aligned ZnO nanowire arrays," *Nanotechnology*, vol. 20, no. 8, p. 5, Feb 2009.

[250] R. A. Laudise and A. A. Ballman, "Hydrothermal synthesis of zinc oxide and zinc sulfide," *J. Phys. Chem.*, vol. 64, no. 5, pp. 688–691, 1960.

[251] M. A. Verges, A. Mifsud, and C. J. Serna, "Formation of rod-like zinc oxide microcrystals in homogeneous solutions," *J. Chem. Soc. Faraday. T.*, vol. 86, no. 6, pp. 959–963, 1990.

[252] L. Vayssieres, K. Keis, S.-E. Lindquist, and A. Hagfeldt, "Purpose-built anisotropic metal oxide material: 3D highly oriented microrod array of ZnO," *J. Phys. Chem. B*, vol. 105, no. 17, pp. 3350–3352, 2001.

[253] J. Nayak, S. N. Sahu, J. Kasuya, and S. Nozaki, "Effect of substrate on the structure and optical properties of ZnO nanorods," *J. Phys. D. Appl. Phys.*, vol. 41, no. 11, p. 6, Jun 2008.

[254] P. C. Chang and J. G. Lu, "ZnO nanowire field-effect transistors," *IEEE. T. Electron. Dev.*, vol. 55, no. 11, pp. 2977–2987, Nov 2008.

[255] S. Xu, *et al.*, "Patterned growth of vertically aligned ZnO nanowire arrays on inorganic substrates at low temperature without catalyst," *J. Am. Chem. Soc.*, vol. 130, no. 45, pp. 14958–14959, 2008.

[256] K. Govender, D. S. Boyle, P. B. Kenway, and P. O'Brien, "Understanding the factors that govern the deposition and morphology of thin films of ZnO from aqueous solution," *J. Mater. Chem.*, vol. 14, no. 16, pp. 2575–2591, 2004.

[257] M. N. R. Ashfold, R. P. Doherty, N. G. Ndifor-Angwafor, D. J. Riley, and Y. Sun, "The kinetics of the hydrothermal growth of ZnO nanostructures," *Thin Solid Films*, vol. 515, no. 24, pp. 8679–8683, Oct 2007.

[258] A. Sugunan, H. C. Warad, M. Boman, and J. Dutta, "Zinc oxide nano-wires in chemical bath on seeded substrates: Role of hexamine," *J. Sol-Gel. Sci. Techn*, vol. 39, no. 1, pp. 49–56, July 01 2006.

[259] S. Xu, C. Lao, B. Weintraub, and Z. L. Wang, "Density-controlled growth of aligned ZnO nanowire arrays by seedless chemical approach on smooth surfaces," *J. Mater. Res.*, vol. 23, no. 8, pp. 2072–2077, Aug 2008.

[260] L. E. Greene, *et al.*, "Low-temperature wafer-scale production of ZnO nanowire arrays," *Angew. Chem. Int. Edit.*, vol. 42, no. 26, pp. 3031–3034, 2003.

[261] H.-H. Park, *et al.*, "Position-controlled hydrothermal growth of ZnO nanorods on arbitrary substrates with a patterned seed layer via ultraviolet-assisted nanoimprint lithography," *Cryst. Eng. Comm.*, vol. 15, no. 17, pp. 3463–3469, 2013.

[262] J. Liu, J. C. She, S. Z. Deng, J. Chen, and N. S. Xu, "Ultrathin seed-layer for tuning density of ZnO nanowire arrays and their field emission

characteristics," *J. Phys. Chem. B*, vol. 112, no. 31, pp. 11685–11690, Aug 2008.

[263] T. Ma, M. Guo, M. Zhang, Y. J. Zhang, and X. D. Wang, "Density-controlled hydrothermal growth of well-aligned ZnO nanorod arrays," *Nanotechnology*, vol. 18, no. 3, p. 7, Jan 2007.

[264] H. Q. Le, S. J. Chua, K. P. Loh, E. A. Fitzgerald, and Y. W. Koh, "Synthesis and optical properties of well aligned ZnO nanorods on GaN by hydrothermal synthesis," *Nanotechnology*, vol. 17, no. 2, pp. 483–488, Jan 2006.

[265] T. Pauporté, D. Lincot, B. Viana, and F. Pellé, "Toward laser emission of epitaxial nanorod arrays of ZnO grown by electrodeposition," *Appl. Phys. Lett*, vol. 89, no. 23, p. 233112, 2006.

[266] M. Law, L. E. Greene, J. C. Johnson, R. Saykally, and P. Yang, "Nanowire dye-sensitized solar cells," *Nat. Mater.*, vol. 4, p. 455, 2005.

[267] C. Xu, P. Shin, L. Cao, and D. Gao, "Preferential growth of long ZnO nanowire array and its application in dye-sensitized solar cells," *J. Phys. Chem. B*, vol. 114, no. 1, pp. 125–129, 2010.

[268] Y. Zhou, W. B. Wu, G. D. Hu, H. T. Wu, and S. G. Cui, "Hydrothermal synthesis of ZnO nanorod arrays with the addition of polyethyleneimine," *Mater. Res. Bull.*, vol. 43, no. 8–9, pp. 2113–2118, 2008.

[269] Z. R. Tian, J. A. Voigt, J. Liu, B. McKenzie, and M. J. McDermott, "Biomimetic arrays of oriented helical ZnO nanorods and columns," *J. Am. Chem. Soc.*, vol. 124, no. 44, pp. 12954–12955, 2002.

[270] L. Xu, Y. Guo, Q. Liao, J. Zhang, and D. Xu, "Morphological control of ZnO nanostructures by electrodeposition," *J. Phys. Chem. B*, vol. 109, no. 28, pp. 13519–13522, 2005.

[271] Z. R. Tian, *et al.*, "Complex and oriented ZnO nanostructures," *Nat. Mater.*, vol. 2, p. 821, 2003.

[272] Y. H. Ni, X. W. Wei, X. Ma, and J. M. Hong, "CTAB assisted one-pot hydrothermal synthesis of columnar hexagonal-shaped ZnO crystals," *J. Cryst. Growth*, vol. 283, no. 1–2, pp. 48–56, Sep 2005.

[273] Y. X. Wang, X. Y. Fan, and J. Sun, "Hydrothermal synthesis of phosphate-mediated ZnO nanosheets," *Mater. Lett.*, vol. 63, no. 3–4, pp. 350–352, 2009.

[274] Y. Sun, L. Wang, X. Yu, and K. Chen, "Facile synthesis of flower-like 3D ZnO superstructures via solution route," *Cryst. Eng. Comm.*, vol. 14, no.9, pp.3199–3204, 2012.

[275] S. H. Ko, *et al.*, "Nanoforest of hydrothermally grown hierarchical ZnO nanowires for a high efficiency dye-sensitized solar cell," *Nano Lett.*, vol. 11, no.2, pp.666–671, 2011.

[276] M. Wanit, *et al.*, "ZnO nano-tree growth study for high efficiency solar cell," *Energy Procedia*, vol.14, pp.1093–1098, 2012.

[277] T. Zhang, W. Dong, M. Keeter-Brewer, S. Konar, R. N. Njabon, and Z. R. Tian, "Site-specific nucleation and growth kinetics in hierarchical nanosyntheses of branched ZnO crystallites," *J. Am. Chem. Soc.*, vol. 128, no. 33, pp. 10960–10968, 2006.

[278] M. R. Alenezi, "Nanostructured zinc oxide sensors," Thesis (Ph.D.) – University of Surrey, 2014.

[279] J. H. Tian, *et al.*, "Improved seedless hydrothermal synthesis of dense and ultralong ZnO nanowires," *Nanotechnology*, vol. 22, no.24, p. 9, May 2011, Art. no. 245601.

[280] J. H. Lee, "Gas sensors using hierarchical and hollow oxide nanostructures: overview," *Sens. Actuator B-Chem.*, vol. 140, no. 1, pp. 319–336, Jun 2009.

[281] T. P. Chou, Q. F. Zhang, G. E. Fryxell, and G. Z. Cao, "Hierarchically structured ZnO film for dye-sensitized solar cells with enhanced energy conversion efficiency," *Adv. Mater.*, vol. 19, no. 18, Sep 2007.

[282] A. M. Cao, *et al.*, "Hierarchically structured cobalt oxide (Co_3O_4): the morphology control and its potential in sensors," *J. Phys. Chem. B*, vol. 110, no. 32, pp. 15858–15863, Aug 2006.

[283] L. Zhang, W. Wang, Z. Chen, L. Zhou, H. Xu, and W. Zhu, "Fabrication of flower-like Bi2WO6 superstructures as high performance visible-light driven photocatalysts," *J. Mater. Chem.*, vol. 17, no. 24, pp. 2526–2532, 2007.

[284] H. Xu, Z. Zheng, L. Z. Zhang, H. L. Zhang, and F. Deng, "Hierarchical chlorine-doped rutile TiO_2 spherical clusters of nanorods: large-scale synthesis and high photocatalytic activity," *J. Solid State Chem.*, vol. 181, no. 9, pp. 2516–2522, Sep 2008.

[285] S. Sun, W. Wang, H. Xu, L. Zhou, M. Shang, and L. Zhang, "$Bi_5FeTi_3O_{15}$ hierarchical microflowers: hydrothermal synthesis, growth mechanism, and associated visible-light-driven photocatalysis," *J. Phys. Chem. B*, vol. 112, no. 46, pp. 17835–17843, 2008.

[286] Y. Qin, X. Wang, and Z. L. Wang, "Microfibre–nanowire hybrid structure for energy scavenging," *Nature*, vol. 451, p. 809, 2008.

[287] H. Yang, L. Hao, N. Zhao, C. Du, and Y. Wang, "Hierarchical porous hydroxyapatite microsphere as drug delivery carrier," *Cryst. Eng. Comm.*, vol. 15, no. 29, pp. 5760–5763, 2013.

[288] Y. Zhang, J. Xu, Q. Xiang, H. Li, Q. Pan, and P. Xu, "Brush-like hierarchical ZnO nanostructures: synthesis, photoluminescence and gas

sensor properties," *J. Phys. Chem. B*, vol. 113, no. 9, pp. 3430–3435, 2009.

[289] J. G. Wen, J. Y. Lao, D. Z. Wang, T. M. Kyaw, Y. L. Foo, and Z. F. Ren, "Self-assembly of semiconducting oxide nanowires, nanorods, and nanoribbons," *Chem. Phys. Lett.*, vol. 372, no. 5, pp. 717–722, 2003.

[290] B. Liu and H. C. Zeng, "Hollow ZnO microspheres with complex nanobuilding units," *Chem. Mater.*, vol. 19, no. 24, pp. 5824–5826, 2007.

[291] P. X. Gao and Z. L. Wang, "Nanopropeller arrays of zinc oxide," *Appl. Phys. Lett.*, vol. 84, no. 15, pp. 2883–2885, 2004.

[292] A. J. Baca, *et al.*, "Semiconductor wires and ribbons for high-performance flexible electronics," *Angew. Chem. Int. Edit.*, vol. 47, no. 30, pp.5524–5542, 2008.

Flexible and Large-Area Electronics

Ravinder Dahiya

University of Glasgow

Ravinder Dahiya is Professor of Electronic and Nanoengineering and an EPSRC Fellow at the University of Glasgow. He is the Director of Electronics Systems Design Centre at the University of Glasgow and leads the multidisciplinary group, Bendable Electronics and Sensing Technologies (BEST). He is President-Elect and a Distinguished Lecturer of the IEEE Sensors Council and serves on the Editorial Boards of the *Scientific Reports, IEEE Sensors Journal*, and *IEEE Transactions on Robotics*. He is an expert in the field of flexible and bendable electronics, robotics and electronic skin.

Luigi G. Occhipinti

University of Cambridge

Luigi G. Occhipinti is Director of Research at the University of Cambridge, Engineering Department, and Deputy Director and COO of the Cambridge Graphene Centre. He is the founder and CEO of Cambridge Innovation Technologies Consulting Limited, providing research and innovation within both the health care and medical fields. He is a recognized expert in printed, organic, and large-area electronics and integrated smart systems with more than twenty years' experience in the semiconductor industry, and is a former R&D Senior Group Manager and Programs Director at STMicroelectronics.

About the Series

This innovative series provides authoritative coverage of the state-of-the-art in bendable and large-area electronics. Specific Elements provide in-depth coverage of key technologies, materials and techniques for the design and manufacturing of flexible electronic circuits and systems, as well as cutting-edge insights into emerging real-world applications. This series is a dynamic reference resource for graduate students, researchers, and practitioners in electrical engineering, physics, chemistry and materials science.

Cambridge Elements $\equiv$

Flexible and Large-Area Electronics

Elements in the Series

Printed in the United States
By Bookmasters